T0192410

Industrial Crystallization of Melts

INDUSTRIAL CRYSTALLIZATION OF MELTS

C. M. VAN 'T LAND
Van 't Land Processing
Enschede, The Netherlands

CRC Press is an imprint of the
Taylor & Francis Group, an **informa** business

First published 2005 by Marcel Dekker.

Published 2020 by CRC Press
Taylor & Francis Group
6000 Broken Sound Parkway NW, Suite 300
Boca Raton, FL 33487-2742

ISBN 13: 978-0-367-57828-2 (pbk)
ISBN 13: 978-0-8247-4111-2 (hbk)

Visit the Taylor & Francis Web site at
http://www.taylorandfrancis.com

and the CRC Press Web site at
http://www.crcpress.com

Library of Congress Cataloging-in-Publication Data
A catalog record for this book is available from the Library of Congress.

Preface

Many bulk products of the chemical industry are present in the solid phase at ambient temperature. Products having a melting point in the temperature range 0–400 °C, however, are in the plant often processed in the liquid phase. In the downstream part of the plant, the liquid is often converted into a solid. The majority of these products are organic compounds, often having melting points in the temperature range 0–200 °C. The greatest density of the latter compounds is at 100 °C.

This book treats the orderly conversion of melts into a solid material to produce a salable product. After the transformation, the product is present as, e.g., flakes, slabs, pastilles, prills, or granules. This book is also concerned with the crystallization of organic melts that do not crystallize in a straightforward manner and need special provisions. Orderly conversion followed by remelting to accomplish purification is also covered.

First, the theoretical background concerning the nature of crystals and instationary heat transfer is discussed. Specific equipment is treated in further chapters. The objective of this book is to assist the process development engineer, the process engineer, and the plant engineer in the selection of equipment for the crystallization and purification of melts. The criteria to be observed are discussed, as are the procedures for sizing equipment. Ample attention is paid to pilot-plant tests and scaling-up.

C. M. van 't Land

Acknowledgments

The writing of this book has been made possible by permission of Akzo Nobel Chemicals, to whose management I am especially grateful. The invaluable experience gained while in their employ has been an important element in the design of this book. Thanks are due to D.J. Buckland of Akzo Nobel Chemicals for improving my English.

Particular appreciation is expressed for the assistance given by:

A. Bouwmeester, Goudsche Machinefabriek: drum flaker
H. Hittenhausen, Akzo Nobel Engineering: caustic soda
G.J. Kruisinga, Akzo Nobel Base Chemicals: prilling
R. Monstrey, Hydro Fertilizer Technology: fluid bed granulation

The author thanks the following companies that most kindly provided data and/or photographs:

Akzo Nobel Base Chemicals, Amersfoort, The Netherlands
APV Baker, Newcastle-under-Lyme, England
BEFS Prokem, Mulhouse, France
Flexsys, Brussels, Belgium
Goudsche Machinefabriek, Waddinxveen, The Netherlands
Hein, Lehmann Trenn- und Fördertechnik, Düsseldorf, Germany
Hydro Fertilizer Technology, Brussels, Belgium

Kaiser, Krefeld, Germany
Koch-Glitsch Italia, Milan, Italy
Kreber, Vlaardingen, The Netherlands
Robatel, Genas, France
Sandvik Process Systems, Fellbach/Stuttgart, Germany
Steel Belt Systems, Milan, Italy
Sulzer Chemtec, Buchs, Switzerland
Theissen, Munich, Germany
Toyo Engineering Corporation, Chiba, Japan
Tsukishima, Kikai, Japan

I wish to thank the following persons and publishers, who most kindly provided permission to use material:

Informations Chimie, Paris, France
L.P.B.M. Janssen, Groningen, The Netherlands
McGraw-Hill Book Company, New York, United States of America
VDI Verlag, Düsseldorf, Germany
VSSD, Delft, The Netherlands

I am greatly indebted to my wife, Annechien, for her patience and constant encouragement.

Contents

1

The Crystalline State of Matter

1.1. INTRODUCTION

This book is on the crystallization of melts. In this chapter, various aspects of the liquid state and the crystalline state of matter are discussed. The majority of melts to be crystallized are organic while some melts, e.g., NH_4NO_3, are inorganic. The crystallization of molten metals and alloys will not be dealt with. All matter is made up of atoms. The forces that hold atoms together are discussed in Sec. 1.2. Ionic, metallic, and covalent bonds are the three primary interatomic forces.

In ionic crystals, positive ions are surrounded by negative ions and vice versa. As a result, the attraction is equal in all directions. An entity like a molecule cannot be distinguished. Metallic crystals are made up of positive ions and "free" electrons. Again, the concept of a molecule is meaningless. In diamond, the carbon atoms are combined into a three-dimensional (3-D) crystal lattice by means of covalent bonds. A molecular entity cannot be distinguished here either. In molecular crystals, the atoms in the molecules are strongly bonded together by covalent bonds. The bonds between the molecules are relatively weak Van der Waals forces. Molecular crystals are softer than the other three types of crystals and have much lower melting points. H_2, H_2O, HNO_3, and the alkanes are examples of materials forming molecular crystals. The characteristics of molecular bonding are discussed in Sec. 1.3.

Crystalline substances are compared to amorphous solids in Sec. 1.4. One criterion is the existence of a reproducible x-ray diffraction pattern. The development of external crystal faces is a different criterion for crystallinity. Crystal systems and crystal lattices are discussed in Sec. 1.5.

Miller indices are used to identify planes of atoms or molecules that exist in a crystal. They are considered in Sec. 1.6. Crystal lattice imperfections are discussed in Sec. 1.7. Organic melts can often be supercooled and it is sometimes possible to obtain a glass below the glass transition temperature. These phenomena are discussed in Sec. 1.8. The crystallization of melts that tend to supercool is treated in Chapter 6. A liquid crystal is a state of matter in between a crystal and a liquid. The material can flow; however, there is some form of molecular orientation in the liquid state. Liquid crystals are discussed briefly in Sec. 1.9. Equipment sizing requires knowledge of relevant physical data. Often, they can be found in handbooks or data banks. In Sec. 1.10, recommendations are given regarding the estimation of values if data cannot be found. Crystallization from the melt is a technique to upgrade organic chemicals. It is different from the crystallization of melts in a number of aspects. This is treated in Sec. 1.11. It is discussed more extensively in Chapter 7. Finally, postcrystallization is dealt with in the last section. Industrially, an exotherm of 4–5 K can sometimes be observed for a material leaving a crystallizer. How to cope with this phenomenon is discussed in Sec. 1.12.

1.2. ATOMIC BONDING

An atom is the smallest entity of an element that has all the properties of that element. The general structure of the atom will be considered shortly in order to develop an insight into the factors governing the properties of materials. The nucleus of an atom is composed of protons and neutrons and it is surrounded by electrons. The hydrogen atom is depicted in Fig. 1.1; it is composed of a proton and an electron. The atoms of other elements may be considered as combinations of protons and neutrons in a nucleus surrounded by electrons. The electrons are present in shells having different energy levels. The maximum number of electrons in a given shell is $2n^2$, where n is the number of the shell. The lithium atom is depicted in Fig. 1.2; the first shell contains the maximum number of two electrons. The second shell contains just one electron while the maximum number for this shell is eight. Neon has eight electrons in this second shell. Neon is a noble gas and is very stable. Generally, the presence of eight electrons in an outer shell confers great

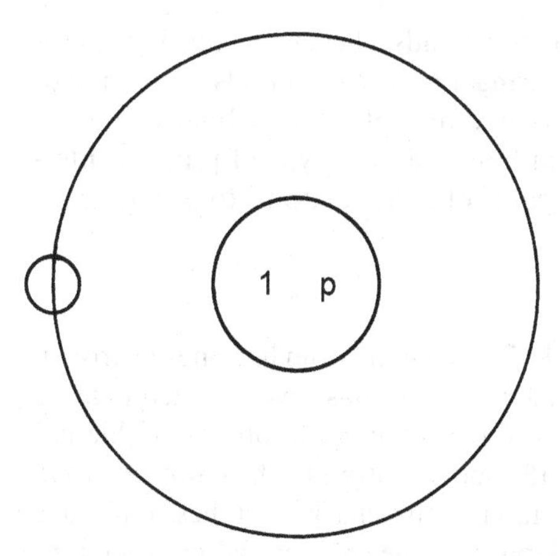

FIG. 1.1 A simplified picture of the hydrogen atom showing one proton in the nucleus and one electron in the first shell.

stability. Atoms try to attain noble gas configuration by one of the following three procedures:

- Receiving extra electrons
- Releasing electrons
- Sharing electrons.

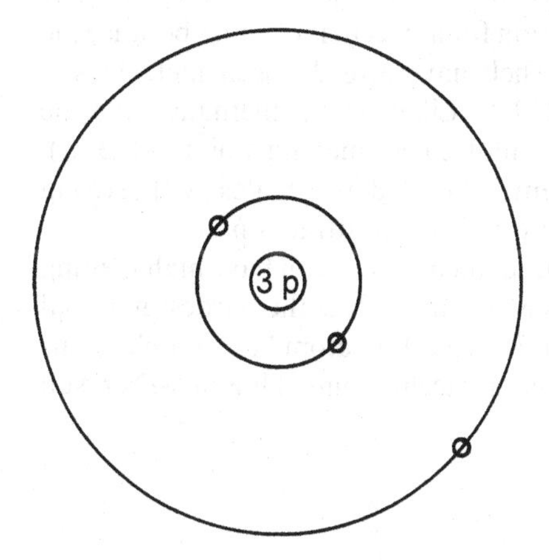

FIG. 1.2 A simplified drawing of the lithium atom showing how a second shell is started because the first one has two electrons.

The first two procedures lead to ionic bonds whereas the third results in a covalent bond. Ionic bonds are strong or primary bonds. Energies of approximately 400 kJ · $gmol^{-1}$ are required to rupture these bonds.

In addition to ionic and covalent bonds, a third type of primary interatomic force, the metallic bond, is capable of holding atoms together.

1.2.1. Ionic Bonds

Solid NaCl is considered as an example. The sodium atom has one electron in its outer (third) shell, whereas the chlorine atom has seven electrons in its outer (third) shell. Both atoms will attain the configuration of a noble gas when an electron is transferred from the sodium atom to the chlorine atom. Ions are produced with a net negative and positive charge and these ions have coulombic attraction to each other. However, the sodium and chloride ions do not join up as pairs. There would be negligible attraction between different pairs if this were the case. A negligible attraction is not in line with, e.g., a melting point of 801 °C. Positive sodium ions surround themselves with negative chloride ions and vice versa. As a result, the attraction is equal in all directions.

1.2.2. Covalent Bonds

Sometimes an atom may acquire two or eight electrons in its outer valence shell by sharing electrons with an adjacent atom. The hydrogen molecule, H_2, is the simplest example. Here, two hydrogen atoms join up as a pair. The resulting molecule has very little attraction for adjacent molecules because the electronic requirements in the outer shell have already been met. This is illustrated by its low melting point: −259 °C. Likewise, the methane molecule acts almost independently from other molecules (melting point −183 °C). Many organic molecules are covalently bonded molecules with strong intramolecular attractions but weak intermolecular attractions.

On the other hand, materials with covalent bonds can have high melting points and great strength. Diamond is an example. It is the hardest material found in nature, and is entirely carbon. Each carbon atom has four electrons in its outer shell, which it shares with four adjacent atoms. Thus, a 3-D crystal lattice is formed.

1.2.3. Metallic Bonds

An atom having only a few valence electrons in its outer shell can lose these electrons relatively easily. The balance of the electrons are held firmly to the nucleus. Thus, a structure of positive ions and "free" electrons is obtained.

The free electrons are also described as an electron "cloud" or "gas." This description provides a useful explanation for many of the properties of metals. The free electrons give the metal its characteristically high thermal conductivity. They transfer thermal energy from a high to a low temperature level. Furthermore, the valence electrons are free to move in an electric field, giving the metal its characteristically high electrical conductivity.

Ionic bonds, covalent bonds, and metallic bonds are relatively strong primary bonds that hold atoms together. Van der Waals forces are weaker secondary bonds supplying interatomic attraction. These forces cause the condensation of noble gases at extremely low temperatures. None of the primary bonds can be effective as the noble gases have a full complement of eight electrons in their outer shells (He: 2). The random movement of the electrons in noble gas atoms causes a momentary polarization. This polarization is called the dispersion effect. The resulting interatomic attractions are weak but real.

1.3. MOLECULAR BONDING

When a limited number of atoms is strongly bonded together while the bonds with other, similar groups of atoms are relatively weak, the resulting structure is called a molecule. Examples are compounds like H_2, H_2O, CO_2, and HNO_3. Usually, the atoms are held together by covalent bonds. However, ionic bonds also occur. The weak intermolecular bonds result from Van der Waals forces. In any asymmetric molecule, the center of positive charge and the center of negative charge do not coincide. An electrical dipole is formed, thus providing a first mechanism for molecular bonding. The molecular polarization of HF is responsible for its high boiling point: 19.4°C. Most diatomic molecules have lower boiling points. Symmetric molecules are bonded together by interatomic attractions caused by the dispersion effect (see Sec. 1.2 on atomic bonding). H_2 and CH_4 are examples of symmetric molecules. In intermolecular bonding, a third type of weak bonding is active, i.e., the hydrogen bond, a special case of molecular polarization. The hydrogen bond is caused by the attraction between a hydrogen nucleus of one molecule and unshared electrons of the oxygen (or nitrogen) of adjacent molecules. This type of intermolecular bonding is responsible for the high boiling temperature and heat of evaporation of water.

The aforementioned molecules are comparatively small, other molecules have larger numbers of atoms. However, the distinction between the strong intramolecular and the weaker intermolecular bonds still holds. The paraffins have C_nH_{2n+2} as their general formula. All intramolecular bonds are covalent bonds. The molecules act individually and have only weak attraction for one another. Their relatively low melting points confirm the weak intermolecular bonds. However, the melting point rises when the molecule becomes larger. Paraffin contains approximately 30 carbon atoms per molecule and is a solid at room temperature. The melting point of C_nH_{2n+2} approaches 145°C asymptotically as n increases.

1.4. CRYSTAL STRUCTURE

Solids may be crystalline or amorphous. In crystals, the constituent molecules, atoms, or ions are arranged regularly into a lattice. Amorphous solids lack such internal regularity.

Sodium chloride is an example of a material forming ionic crystals. Ionic bonds are strong and this explains its high melting point: 801°C.

In diamond, the carbon atoms are covalently bonded. As a result, atomic crystals are obtained. The bonds between the atoms are very strong as is illustrated by diamond's great hardness (10 on Mohs' scale). In metals, the metallic bond holds the atoms together. Like atoms and ions, molecules can form a crystalline pattern. However, molecular crystals are distinctly different from both ionic and atomic crystals. Unlike atoms and ions, molecules are not spherical. Furthermore, the intramolecular bonds (often covalent bonds) are strong while the intermolecular attractions are weak Van der Waals forces. Many organic melts have melting points in the range 50–150°C. Often, the heats of fusion are in the range 100–200 $kJ \cdot kg^{-1}$. It is interesting to note that cyclohexane has a very low heat of fusion: 31.7 $kJ \cdot kg^{-1}$. Its melting point is 6.7°C. In cyclohexane, all bonds are single pairs of covalent electrons. Each carbon atom is surrounded by a full complement of four neighboring atoms. Here, the intramolecular bonds are quite strong while the intermolecular attractions are very weak. In the process industry, melts are crystallized into molecular crystals (e.g., naphthalene) and ionic crystals (e.g., NH_4NO_3).

The presence of a lattice structure can be detected by means of x-ray analyses. X-rays are diffracted by the planes of atoms, ions, or molecules within the crystal when they are directed at a crystalline material. By x-ray analysis, many of the substances once considered amorphous have now been shown to be crystalline. However, the development of external crystal faces is frequently seen as a proper criterion for crystallinity. Some substances look like solids; however, they flow under pressure and are hence viscous liquids. Glasses will be discussed separately in Sec. 1.8.

1.5. CRYSTAL SYSTEMS AND CRYSTAL LATTICES

The elementary particles of a crystalline material (atoms, ions, or molecules) may be packed according to one of seven main crystal systems. These seven crystal systems are related to the way in which space can be divided into equal volumes by intersecting plane surfaces. The first six of these systems can be described by means of three axes, x, y, and z. The three axes and the angles between the axes are shown in Fig. 1.3. The description of the seventh system, the hexagonal system, requires four axes. Here, as usual, the z-axis is vertical and this axis is perpendicular to the other three axes (x, y, and u). The latter three are inclined at 60° to one another. The seven crystal systems are shown in Table 1.1. The cubic system is the simplest and most regular system. Fig. 1.4 depicts the simple cubic structure (the body-centered and face-centered

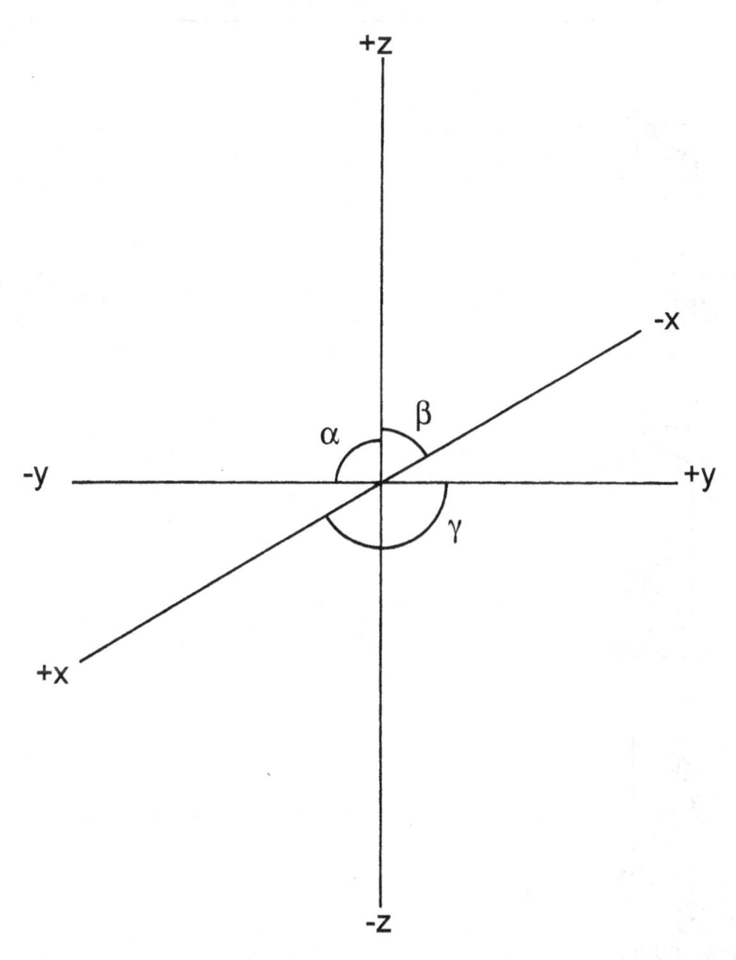

FIG. 1.3 Crystallographic axes for describing the seven crystal systems but for the hexagonal system.

TABLE 1.1 The Seven Crystal Systems

System	Axial dimensions	Axial angles	Example
Cubic	$x = y = z$	$\alpha = \beta = \gamma = 90°$	NaCl
Tetragonal	$x = y \neq z$	$\alpha = \beta = \gamma = 90°$	$NiSO_4 \cdot 7H_2O$
Orthorhombic	$x \neq y \neq z$	$\alpha = \beta = \gamma = 90°$	α-S
Monoclinic	$x \neq y \neq z$	$\alpha = \beta = 90° \neq \gamma$	Oxalic acid
Triclinic	$x \neq y \neq z$	$\alpha \neq \beta \neq \gamma \neq 90°$	$CuSO_4 \cdot 5H_2O$
Trigonal	$x = y = z$	$\alpha = \beta = \gamma \neq 90°$	$NaNO_3$
Hexagonal	$x = y = u \neq z$	x, y, u at 60°; z at 90° with x, y, u	Ice

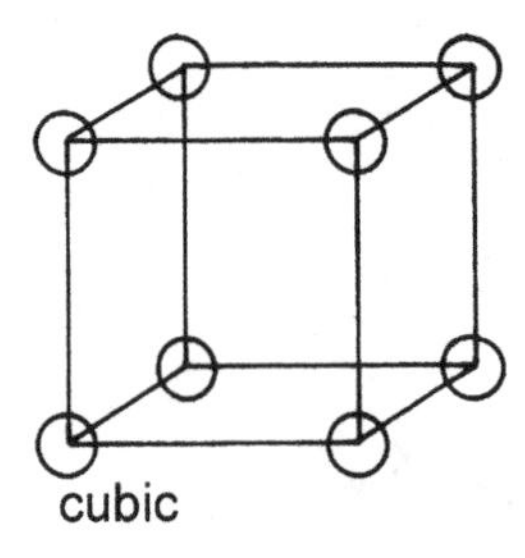

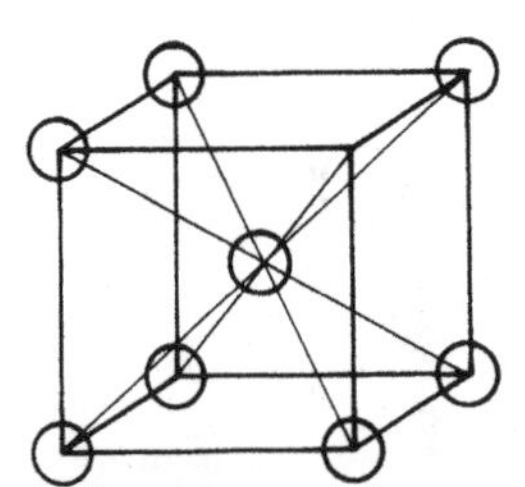

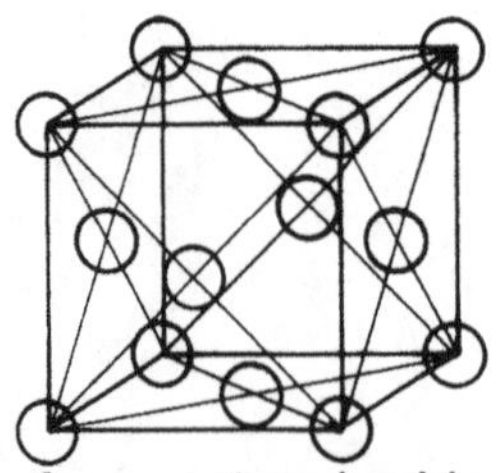

FIG. 1.4 The three cubic lattices.

TABLE 1.2 The Fourteen Space Lattices

Type of symmetry	Lattice
Cubic	Cube
	Body-centered cube
	Face-centered cube
Tetragonal	Square prism
	Body-centered square prism
Orthorhombic	Rectangular prism
	Body-centered rectangular prism
	Rhombic prism
	Body-centered rhombic prism
Monoclinic	Monoclinic parallelepiped
	Clinorhombic prism
Triclinic	Triclinic parallelepiped
Rhomboidal	Rhombohedron
Hexagonal	Hexagonal prism

structure are also shown). For practical purposes, it is possible to subdivide the seven crystal systems into 14 crystal lattices. All crystals have space lattices that fall into one of the 14 categories shown in Table 1.2.

The shape of a grown crystal does not always reflect the symmetry expected from its basic unit cell. The reason is that the most commonly occurring faces are those that are parallel to lattice planes having the highest density of points.

There is a relationship between the lattice bonding and the crystal system a substance crystallizes in. Usually, the cubic or hexagonal system is favored by crystalline elements and simple inorganic compounds because these systems allow efficient packing. About 80% of the known crystalline organic substances and 60% of the natural minerals crystallize in orthorombic or monoclinic systems (Mullin, 1993).

1.6. MILLER INDICES

Miller indices are used to identify planes of atoms or molecules that exist in a crystal. These (*hkl*) symbols are the coordinates of vectors perpendicular to these planes. For this characterization, the unit cell is used as a basis. Three directions within an orthorhombic lattice are drawn in Fig. 1.5, i.e., the [111], [110], and [010] directions. The (111), (110), and (010) planes are perpendicular to the corresponding vectors. A [424] direction is identical to a [212] direction, so the lowest combination of integers is used. If a minus sign is shown above the digit, the Miller index is negative, e.g., $(\bar{1}1\bar{1})$. In terms of cell

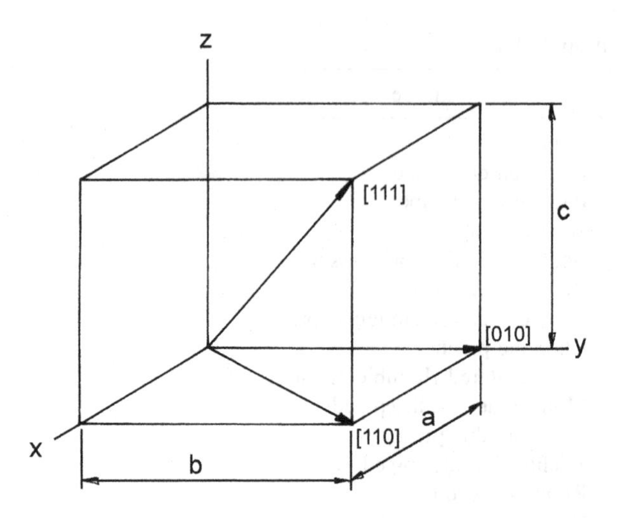

FIG. 1.5 Crystal directions.

unit distances from the origin, the numbers used are the reciprocals of the intercepts on the axes. For example, the (010) plane cuts the *y*-axis at 1, while the plane is parallel to the *x*- and *z*-axes:

$$\frac{1}{\infty}, \frac{1}{1}, \frac{1}{\infty} = (010)$$

1.7. IMPERFECTIONS

Lattice imperfections can be found in most crystals. Point defects, line defects, and surface imperfections can be distinguished.

1.7.1. Point Defects

The absence of an elementary particle from a normally occupied site is called a vacancy. A substitutional impurity particle is a foreign particle in the site of a matrix particle, and an interstitial impurity particle is a foreign particle in an interstice between matrix particles. A matrix particle can also be self-interstitial.

A particle in a crystal may leave its site and dissolve interstitially in the structure. The resulting vacancy and interstitial atom is called a Frenkel imperfection. A Schottky imperfection is specific for an ionic solid. It comprises an anion vacancy and an associated cation vacancy. Schottky imperfections are encountered more often than Frenkel imperfections, because few structures have interstices large enough to accommodate particles.

Point defects have not exhaustively been discussed.

1.7.2. Line Defects

Dislocations are the most common type of line imperfections. An edge dislocation may be described as an edge of an extra plane of elementary particles within a crystal structure (see Fig. 1.6). Edge dislocations can be caused by a small mismatch in the orientation of neighboring parts of a growing crystal. Shear can also introduce this type of dislocation.

A screw dislocation is depicted in Fig. 1.7. This type of imperfection often develops on growing crystals and unit cells can be added to the "step" of the screw. Furthermore, this dislocation can also be caused by shear.

1.7.3. Surface Imperfections

These defects arise from a change in the stacking of elementary particles across a boundary. Grain boundaries of separate crystals having different orientations in a polycrystalline aggregate are examples. There are further surface imperfections.

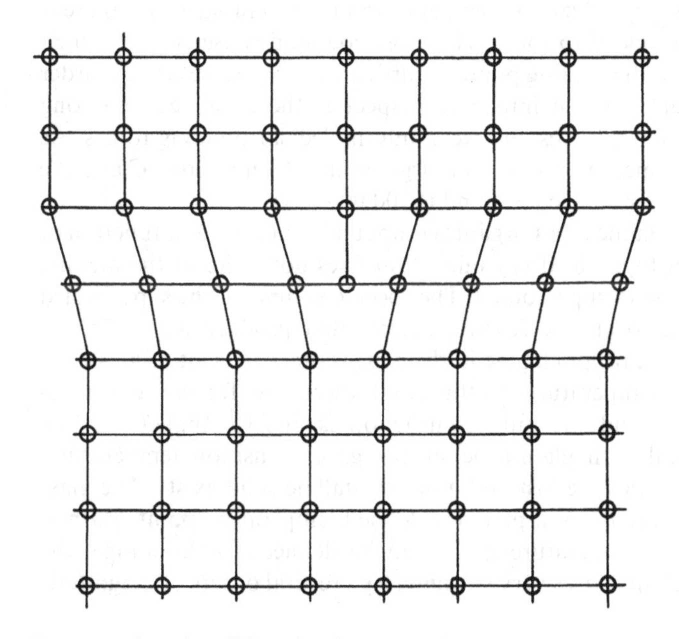

FIG. 1.6 An edge dislocation in a crystal.

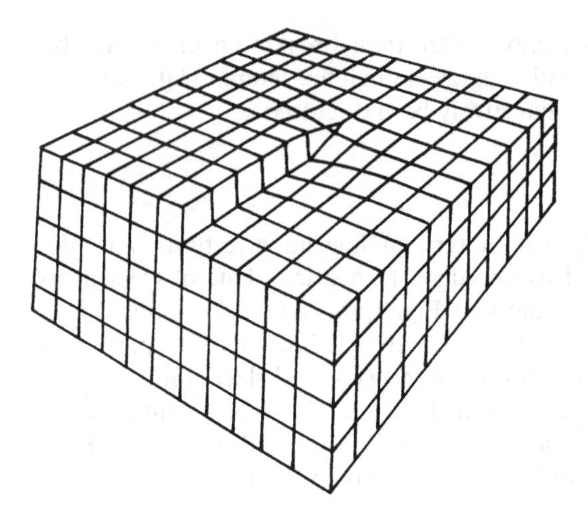

FIG. 1.7 A screw dislocation.

1.8. GLASS TEMPERATURE

Many organic liquids can be made to solidify in the glassy state if they are cooled through the crystallization temperature range so rapidly that there is no time for crystal nuclei to form. The most common cause is a high melt viscosity at or below the melting point. A further possible cause is a low order of molecular symmetry. An important aspect is the existence of strong intramolecular bonding forces and weak intermolecular bonding forces. As rules of thumb, the melting point of an organic compound is 100°C and the heat of fusion of an organic compound is 100,000 $J \cdot kg^{-1}$.

The specific volume of an organic compound is plotted as a function of the temperature in Fig. 1.8. If crystallization does not occur at the melting point, the melt becomes supercooled. The specific volume of the supercooled melt decreases due to atomic rearrangements that produce more efficient packing. There is an abrupt change in the expansion coefficient of glasses at the glass transition temperature. At this point, there is no further rearrangement of the atoms and the only contraction is due to smaller thermal vibrations. We deal with glasses below the glass transition temperature. The fluid characteristics are lost and a noncrystalline solid exists. The glass transition temperature is not a precisely defined temperature point. Rather, the glass transition temperature range can be defined. In this range, the viscosity of the melt increases very strongly, e.g., several orders of magnitude within 10 K.

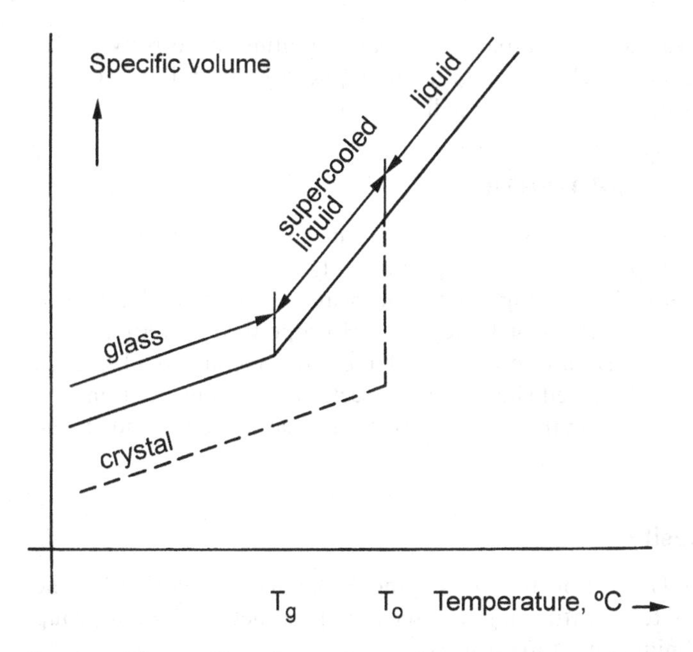

FIG. 1.8 The specific volume of crystals and glasses as a function of temperature.

1.9. LIQUID CRYSTALS

Liquids are usually isotropic, i.e., the properties of a liquid are not dependent on the direction in which the liquid properties are measured. There are liquids, however, that exhibit anisotropy at temperatures just above their melting point. These liquids bear the name "liquid crystals." The anisotropy can be explained by some form of molecular orientation, comparable to, e.g., a number of logs floating down a river. Liquids exhibiting anisotropy are turbid. On heating to temperatures well above the melting point, the liquids become clear at a specific temperature. They become true liquids. Anisotropic liquids can conveniently be divided into three classes:

- The smectic (soaplike) state is characterized by thin layers gliding over one another.
- The nematic (threadlike) state often exhibits mobile threads in the liquid.
- The cholesteric state exhibits strong optical activity.

The name of the latter class of compounds arises from the fact that cholesteryl compounds form the majority of known examples.

Para-Azoxyanisole is a typical material exhibiting anisotropy in the liquid state. The solid melts at 84°C to the liquid crystal and at 150°C it undergoes a transition to the isotropic liquid state.

1.10. PHYSICAL PROPERTIES

Regarding the crystallization of melts, important physical properties of the two phases are the specific mass, the specific heat, and the thermal conductivity. Furthermore, the melting point, the heat of fusion, and the melt viscosity are relevant. Values of these physical properties are required for calculations. Often, they can be found in handbooks or data banks. Each physical property is discussed shortly in this section. Recommendations are given concerning the estimation of values if data cannot be found in the literature.

1.10.1. Viscosity

Poling et al. (2001) mention the Orrick and Erbar method (1974) for the estimation of low-temperature liquid viscosity. This method uses a group contribution technique to estimate A and B:

$$^{e}\log\frac{\mu}{\rho_l M} = A + \frac{B}{T} \tag{1.1}$$

μ	liquid viscosity, cP
ρ_l	liquid specific mass at 20°C, $g \cdot cm^{-3}$
M	molecular weight, $g \cdot gmol^{-1}$
T	temperature, K

Example

It is required to calculate the viscosity of liquid 1-butanol at 120°C. The liquid is under pressure as the boiling point is 117.7°C at atmospheric pressure. The experimental value is 0.394 cP.

$A = -10.79$
$B = 2271$
ρ_l at 20°C $= 0.8096\ g \cdot cm^{-3}$
$M = 74.123\ g \cdot gmol^{-1}$

$$^{e}\log \frac{\mu}{0.8096 \cdot 74.123} = -10.79 + \frac{2271}{393}$$

$$\mu = 0.400 \text{ cP at } 120\,^{\circ}\text{C}$$

1.10.2. Thermal Conductivity

Pure Liquids

Most common organic liquids have values of the thermal conductivity in the range 0.1–0.17 $W \cdot m^{-1} \cdot K^{-1}$ at temperatures below the normal boiling point. Pohling et al. (2001) recommend two empirical estimation techniques for the thermal conductivity of pure liquids. The errors are usually less than 10%.

Solids

The thermal conductivities of solid organic materials are often in the range 0.1–0.2 $W \cdot m^{-1} \cdot K^{-1}$. The thermal conductivity of ice is 3.5 $W \cdot m^{-1} \cdot K^{-1}$ at −100°C and 2.2 $W \cdot m^{-1} \cdot K^{-1}$ at 0°C. Not many thermal conductivities of solid organic materials can be found in the literature. Estimation techniques are not readily available. Carrying out measurements is often necessary.

1.10.3. Melting Point and Heat of Fusion

Matsuoka and Fukushima (1986) looked at the melting points of approximately 10,600 organic compounds. They used the CRC Handbook (Lide, 1976). It was found that 72% of the organic compounds have melting points in the temperature range 0–200°C. The greatest density of compounds is at 100°C.

The authors also found that the heats of fusion are closely related to the structure of the materials, e.g., on plotting the heat of fusion of the n- and i-alkanes in $kJ \cdot gmol^{-1}$ as a function of the melting point, a curve was obtained. A high heat of fusion corresponded with a high melting point and vice versa. The same tendency was found for aromatics. However, the slope of this curve was less steep.

1.10.4. Specific Mass

Pure Liquids

The specific mass of a liquid is significantly temperature dependent. Many organic liquids have a specific mass in the range 500–1000 $kg \cdot m^{-3}$. There are several techniques for estimating pure liquid specific masses (Pohling et al.; 2001). The authors recommend two techniques to estimate saturated liquid specific masses (i.e., at the saturated vapor pressure). The pure component parameters for the Rackett equation (1970) for over 450 compounds are listed in this book. The authors also recommend two techniques for compressed liquids. Directions are given on how to proceed if data are not available for a compound.

Solids

The specific mass of a solid is only slightly temperature dependent. Many organic crystalline solids have a specific mass in the range 1000–2000 $kg \cdot m^{-3}$. In many instances, the specific mass of a solid substance can be found in physical property handbooks or data banks. The calculation of the theoretical specific mass of a crystalline material from the lattice parameters is a feasible step. The reader is referred to books on, e.g., solid state physics.

1.10.5. Specific Heat

Pure Liquids

Many organic liquids have specific heat values in the range 1700–2500 $J \cdot kg^{-1} \cdot K^{-1}$ at ambient temperature. The specific heat of a liquid usually increases with increasing temperature. For example, the values for the heat capacity of benzene at 20° and 60°C are approximately 1700 and 1800 $J \cdot kg^{-1} \cdot K^{-1}$, respectively. The specific heat of water is almost constant in the range 0–100°C (4200 $J \cdot kg^{-1} \cdot K^{-1}$).

Pohling et al. (2001) discuss group contribution methods and corresponding states methods to estimate liquid heat capacities. They recommend use of the Rùzicka–Domalski (1993) method (a group contribution method) below the boiling point. At higher temperatures, they recommend their own method (a corresponding state method).

Solids

Inorganic solids often have specific heat values in the range 400–1300 $J \cdot kg^{-1} \cdot K^{-1}$. Many organic solids have specific heats in the range 800–2100 $J \cdot kg^{-1} \cdot K^{-1}$. Generally, the heat capacity increases slightly with an increase in temperature. To estimate heat capacities of solid substances, two empirical rules are useful. Dulong and Petit's rule states that the atomic heat is 6.2 $cal \cdot gmol^{-1} \cdot K^{-1}$. Kopp's law states that the molar heat capacity equals the sum of the atomic heats of the constituent atoms. On applying the latter rule, it is to be noted that there are exceptions to the approximate value of 6.2 cal · $gmol^{-1}$ · K^{-1}:

C = 1.8	H = 2.3	B = 2.7	Si = 3.8
O = 4.0	F = 5.0	S = 5.4	[H_2O] = 9.8

[H_2O] stands for water as ice or as water of crystallization in solid substances.

For example, to calculate the specific heat of sodium sulfate (Na_2SO_4):

Element	Atomic weight	Atomic heat capacity
Na_2	46.00	12.4
S	32.06	5.4
O_4	64.00 +	16.0 +
Total	142.06	33.8

Specific heat 33.8/142.06 = 0.24 cal · g^{-1} · K^{-1} = 999 J · kg^{-1} · K^{-1}. *Perry's Chemical Engineers' Handbook* (Perry et al., 1997) gives 967 J · kg^{-1} · K^{-1}, i.e., a 3% error from the value derived from Kopp's Law.

Physical Constants: Variation with Temperature

Physical constant	Falls when the temperature rises
Specific mass, liquids	Yes
Specific mass, solids	Yes
Specific heat, liquids	No
Specific heat, solids	No
Thermal conductivity, liquids	Yes
Thermal conductivity, solids	Yes
Viscosity, liquids	Yes

1.11. CRYSTALLIZATION FROM THE MELT

Crystallization from the melt is a unit operation to upgrade organic chemicals. This will be illustrated by means of an example. In a typical plant, the liquid reaction product at 60°C has an assay of 85% by weight. The melting point of the pure product is 40°C. The product must be purified up to 99% by weight in order to obtain a salable material. This is done by cooling crystallization from a solution of the organic chemical in a solvent. A simple experiment shows the feasibility of melt crystallization technology. A sample of the liquid reaction product at 60°C in a plastic bottle is put in a refrigerator at 10°C. After overnight cooling, 80% by weight of the contents of the bottle has crystallized and 20% is still present as liquid. The liquid can be poured out of the bottle, and the crystalline material can be remolten. The assay of the latter liquid is 97% by weight. Thus, purification from 85% up to 97% by weight with a yield of 91% has occurred. First, this principle can be applied more professionally than at the bottle test just described. This will lead to greater purity and higher yield. Second, this principle can be repeated. Then, the impure mother liquor and the purified product move countercurrently through the melt crystallization plant. The separative action of crystallization is due to the selectivity by which atoms, ions, and molecules are built into a crystal lattice. As a separation process, melt crystallization technology is energy efficient. On comparing crystallization from the melt to distillation, it becomes apparent that the heat of fusion of organic chemicals is two to five times smaller than the heat of evaporation. However, the melt crystallization processes applied in the chemical industry show a limited separation efficiency. The operation often has to be repeated.

Furthermore, for a number of chemicals of interest, the process temperature is between 30° and 70°C. This is an attractive temperature range for the utilization of waste heat. Crystallization from the melt is ecologically sound because use of a solvent is avoided.

Crystallization from the melt will be discussed in more detail in Chapter 7. Layer and suspension techniques are used. Common examples of organic chemicals that are isolated by crystallization from the melt on an industrial scale are *para*-xylene, *para*-dichlorobenzene, monochloroacetic acid, and naphthalene (Arkenbout, 1995). Monochloroacetic acid was the first material purified commercially by means of melt crystallization technology. Hoechst has used a simple heat exchanger containing a number of vertical pipes since 1901 to purify the compound for indigo synthesis (Arkenbout, 1995). BEFS Prokem, a French equipment manufacturer, started the development of the PROABD refiner after the observation that naphthalene always has a higher melting point near the wall of a drum than in the middle.

1.12. POSTCRYSTALLIZATION

Sometimes, a product leaving a crystallization device is not fully crystallized. This can be checked by storing a sample in a Dewar vessel and recording the sample's temperature as a function of time. As a rule, postcrystallization is exothermic. The degree of postcrystallization, x, can be approximated as follows:

$$xi = (1 - x)c_s\Delta T \qquad \mathrm{J \cdot kg^{-1}} \tag{1.2}$$

A material exhibiting postcrystallization may be sticky and storing such a material can give caking problems. Bringing the crystallization process to completion in the crystallization device would mean long residence times. The answer often chosen is to subject a shallow layer of product to a ripening operation on, e.g., a conveyor belt.

Example

$i = 100{,}000\ \mathrm{J \cdot kg^{-1}}$
$c_s = 1500\ \mathrm{J \cdot kg^{-1} \cdot K^{-1}}$
$\Delta T = 4\ \mathrm{K}$
$x \cdot 100{,}000 = (1 - x)\ 1500 \cdot 4$
$x = 0.057$ (5.7% liquid by weight)

The conclusion is that this solid, exhibiting an exotherm of 4 K, still contains 5.7% liquid by weight initially.

LIST OF SYMBOLS

A	Constant in Eq. (1.1) [cP]
B	Constant in Eq. (1.1) [cP · K]
c_s	Solid specific heat [$\mathrm{J \cdot kg^{-1} \cdot K^{-1}}$]
i	Heat of fusion [$\mathrm{J \cdot kg^{-1}}$]
M	Molecular weight [$\mathrm{g \cdot gmol^{-1}}$]
T	Temperature [K]
T_g	Glass transition temperature [°C]
T_o	Melting point [°C]
ΔT	Temperature increase [K]
x	Degree of postcrystallization
μ	Liquid dynamic viscosity [cP]
ρ_l	Liquid specific mass at 20°C [$\mathrm{g \cdot cm^{-3}}$]

REFERENCES

Arkenbout, G. F. (1995). *Melt Crystallization Technology*. Lancaster, PA: Technomic.

Lide, D. R. (1976). *CRC Handbook of Chemistry and Physics*. Boca Raton, FL: CRC Press.

Matsuoka, M., Fukushima, H. (1986). Determination of solid liquid equilibrium. *Bunri Gijutsu* 16:4. In Japanese.
Mullin, J. W. (1993). *Crystallization.* Oxford, England: Butterworth.
Perry, R. H., Green, D. W., Maloney, J. O. (1997). *Perry's Chemical Engineers' Handbook.* New York: McGraw-Hill.
Poling, B. E., Prausnitz, J. M., O'Connell, J. P. (2001). *The Properties of Gases and Liquids.* New York: McGraw-Hill.

2

Crystallization of a Layer on a Cooled Surface

2.1. INTRODUCTION

The theoretical foundations of the crystallization and cooling of layers on a cooled surface are laid down in this chapter. The considerations are applied to the processes occurring on a belt cooler and on a drum flaker. The developed theories apply for materials crystallizing readily, i.e., the rate of crystallization is determined by the rate of heat transfer. Experimental results confirming the applicability of the theoretical expressions for solidification on cooling belts and drum flakers will be discussed in Chapters 3 and 4. It is assumed that the cooling of the crystallized layers follows the laws of instationary heat transfer. Sec. 2.2 is a short section describing the solidification process. Gregorig's (1976) approach to describe the crystallization of a layer on a cooled surface is chosen. The model is explained in Sec. 2.3. It is applied to the solidification on a metal wall with a constant temperature in Sec. 2.4. Three different cases are distinguished and the mathematical approaches are relatively simple. Solidification on a metal wall with a variable temperature is attempted in Sec. 2.5. Again, three different cases are distinguished. The mathematical approaches are less simple. However, the definition of four dimensionless numbers is helpful. In Sec. 2.6, the concept of the limiting layer thickness is explained. Finally, the cooling of crystallized layers on a metal wall is discussed in Sec. 2.7.

2.2. THE SOLIDIFICATION PROCESS

Fig. 2.1 illustrates a typical solidification process. The melt, the solid layer, the metal wall, and the cooling medium can be distinguished from left to right. The temperatures of the cooling medium and the metal wall are lower than the melting point of the liquid to be solidified. Due to mixing, the melt has a uniform temperature T_f. The temperature at the melt/solid interface is T_o. The temperature decreases from T_f to T_o across a boundary layer. Likewise, the temperature decreases from T_o to T_w across the solid layer and from T_w to T_i across the metal wall. Finally, the cooling medium has a uniform temperature T_c. The layer thickness δ increases with time. The calculation of the solidification time as a function of δ will be discussed in the sections to follow. Solidification is caused by unsteady state heat transfer. First, the layer thickness changes with time. Second, the metal wall temperature T_w changes with time as well. The basic differential equation for the process is the Fourier equation:

$$\frac{\partial T}{\partial t} = a\frac{\partial^2 T}{\partial \delta^2} \tag{2.1}$$

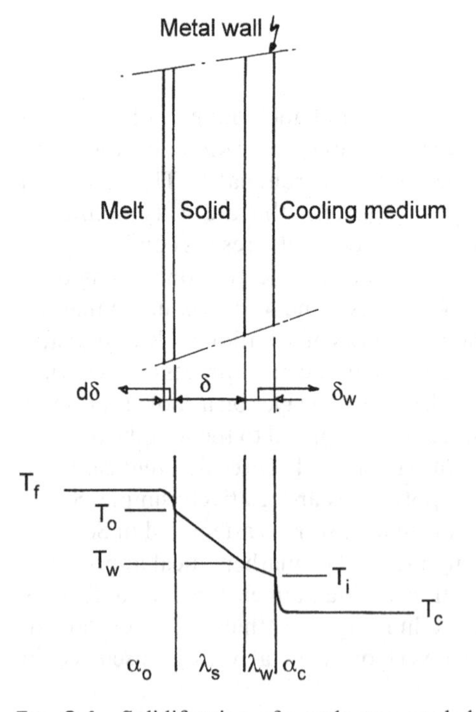

FIG. 2.1 Solidification of a melt on a cooled surface.

This equation is applicable for the unidirectional flow of heat in a continuous medium. The equation must be integrated using boundary conditions.

2.3. GREGORIG'S APPROACH

Gregorig's (1976) approach assumes linear T/δ relationships in the deposited layer and in the metal wall. According to theory, the T/δ lines cannot be straight. However, the relations obtained give results that match experimental results reasonably well. Heertjes and Ong Tjing Gie (1960) discuss solutions of the Fourier equation for three different sets of boundary conditions. The enthalpy change of the deposited solid layer with time was neglected which can be justified by the relatively small temperature effects that occurred in their experiments. The two authors arrived at solutions by introducing simplifications. Their relations are comparable to the relations arrived at on using Gregorig's approach. However, the latter takes the enthalpy change of the product layer into account. A further aspect is that Heertjes' and Ong Tjing Gie's plots of measured T/δ pairs are straight lines. Griesser (1973) calculates in a worked example that the T/δ line is practically straight after 2.25 sec. He discusses the solidification process on drum flakers. It should be stated that the temperature of the wall in his example is not constant.

Gel'perin et al. (1978) summarize the work of various authors. The Fourier equation is the starting point. Gregorig's (1976) equation is a heat balance for the growing layer. See Fig. 2.1. Generally, the heat balance for the growing layer is:

$$\text{heat removal} = \text{heat supply} + \text{depletion} \ \mathrm{W \cdot m^{-2}}$$

Depletion is explained as follows. Material already crystallized cools down as the layer grows. Thus, heat is extracted from the layer as it grows. This is illustrated in Fig. 2.2. Point A cools down by ΔT when the layer thickness has doubled. This is generally called the enthalpy effect.

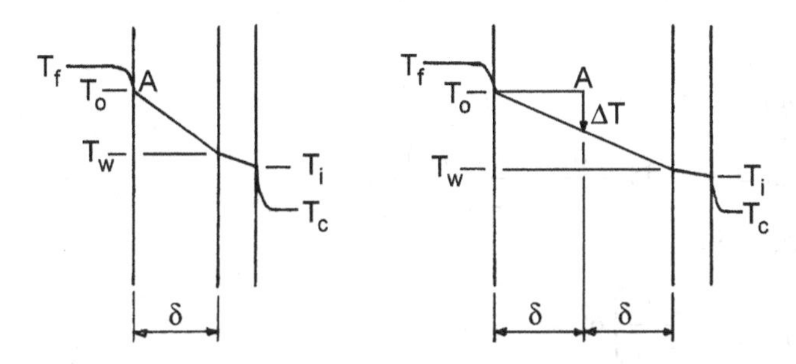

FIG. 2.2 The enthalpy effect.

Gregorig's equation:

$$\frac{\lambda_s(T_o - T_w)}{\delta} \cdot dt = \alpha_o(T_f - T_o)dt + i\rho_s \cdot d\delta + \frac{\rho_s c_s d\{(T_o - T_w)\delta\}}{2} \quad (2.2)$$

This approach does not neglect the enthalpy change of the solid layer with time. This aspect is taken care of by the last part of the right-hand side of the equation. Several theoretical derivations do not take this effect into account, and this can lead to substantial deviations from experimental results. Equation (2.2) can be integrated between ($t = 0, \delta = 0$) and ($t = t, \delta = \delta$). It is assumed that the physical properties do not depend on the temperature.

2.4. SOLIDIFICATION ON A WALL WITH A CONSTANT TEMPERATURE

Reference is made to Fig. 2.1. The resistances to heat transfer are located in:

- The thermal boundary layer melt/solid
- The solid
- The metal wall
- The thermal boundary layer wall/cooling medium.

Solidification on a wall with a constant temperature involves a situation in which the latter two resistances to heat transfer are negligible. In that case, $T_w \equiv T_c$. For example, one pastille is deposited on an aluminum plate, having a thickness of 20 mm, and is cooled by flashing propylene.

Three different cases can be distinguished:

- Case 1 concerns the situation in which the melt is at its melting point.
- Case 2 concerns the situation in which the melt is at a temperature higher than the melting point while the liquid layer is stagnant and the liquid thermal conductivity is zero.
- Case 3 addresses solidification at which the melt is at a temperature exceeding the melting point while heat is transferred convectively from the melt to the deposited layer.

Case 1

Equation (2.2) becomes:

$$\frac{\lambda_s(T_o - T_w)}{\delta} \cdot dt = i\rho_s \cdot d\delta + \frac{\rho_s c_s(T_o - T_w)}{2} \cdot d\delta$$

Integration results in:

$$t = \left\{\frac{i\rho_s}{2(T_o - T_w)\lambda_s} + \frac{\rho_s c_s}{4\lambda_s}\right\}\delta^2 \text{ sec} \quad (2.3)$$

It is also possible to write:

$$\delta = \frac{\sqrt{t}}{\sqrt{\frac{i\rho_s}{2(T_o - T_w)\lambda_s} + \frac{\rho_s c_s}{4\lambda_s}}} \text{ m}$$

Equation (2.3) states that the solidification time is a quadratic function of the layer thickness. The deceleration of the growth rate of the deposited layer is caused by the increasing resistance of the growing layer to heat transfer. Basically, the differential equation has the form: $d\delta/dt \sim 1/\delta$. In words: the rate of growth of the layer is inversely proportional to the layer thickness at any point in time.

A parallel can be drawn with the well-known equation of motion: $s = 0.5at^2$. This equation is derived from: $ds/dt \sim t$.

In many instances, experimental observations approximately confirm this "quadratic growth law."

For example, Preger (1968, 1970, 1980) states that this quadratic growth law approximately describes his experimental drum flaker results.

He states:

$$\delta = An^{(z-1)} \tag{2.4}$$

This is a power law equation in which z is approximately 0.5 (0.4–0.6).

It follows:

$$\delta \approx \frac{A}{n^{0.5}} \quad \text{and} \quad \delta \approx Bt^{0.5}$$

The latter equation has the same form as Eq. (2.3). For more details, see Sec. (2.5).

Case 2

Equation (2.2) becomes:

$$\frac{\lambda_s(T_o - T_w)}{\delta} \cdot dt = \{i + c_l(T_f - T_o)\}\rho_s \cdot d\delta + \frac{\rho_s c_s(T_o - T_w)}{2} \cdot d\delta$$

Integration results in:

$$t = \left[\frac{\{i + c_l(T_f - T_o)\}\rho_s}{2(T_o - T_w)\lambda_s} + \frac{\rho_s c_s}{4\lambda_s}\right]\delta^2 \text{ sec} \tag{2.5}$$

The solidification is abstracted to the simultaneous transfer of the sensible heat and the heat of fusion as the solidification front progresses. Preger (1969 Aufbereitungs-Technik) suggests this approach. The background is that usually the melt has a temperature that exceeds the melting point by 5, 10, or 15 K only. Thus, an allowance for the sensible heat is required but an inaccuracy in this allowance is reflected in the final result only marginally.

Equation (2.5) also describes the quadratic dependence of the time on the layer thickness, however, the coefficient of δ^2 is different from the coefficient of δ^2 in Eq. (2.3).

Case 3

Equation (2.2):

$$\frac{\lambda_s(T_o - T_w)}{\delta} \cdot dt = \alpha_o(T_f - T_o)dt + i\rho_s \cdot d\delta + \frac{\rho_s c_s(T_o - T_w)}{2} \cdot d\delta$$

This equation can be simplified to:

$$dt - C_1\delta \cdot dt = C_2\delta \cdot d\delta$$

$$C_1 = \frac{\alpha_o(T_f - T_o)}{\lambda_s(T_o - T_w)}, \quad C_2 = \frac{2i\rho_s + \rho_s c_s(T_o - T_w)}{2\lambda_s(T_o - T_w)}$$

$$dt = \frac{C_2\delta \cdot d\delta}{(1 - C_1\delta)} \quad \text{and} \quad \int dt = C_2 \int \frac{\delta \cdot d\delta}{(1 - C_1\delta)}$$

The integral can be solved by putting $1 - C_1\delta = z$.

Thus, $\delta = (1 - z)/C_1$ and $d\delta = -dz/C_1$. The integration occurs between $z = 1$ and $z = 1 - C_1\delta$. The result is:

$$t = \frac{C_2}{C_1^2}\{-C_1\delta - {}^e\log(1 - C_1\delta)\}$$

Substitution of C_1 and C_2 leads to:

$$\begin{aligned} t = &-\frac{\lambda_s(T_o - T_w)\{2i\rho_s + \rho_s c_s(T_o - T_w)\}}{2\alpha_o^2(T_f - T_o)^2} \\ &\times {}^e\log\left\{\frac{\lambda_s(T_o - T_w) - \alpha_o(T_f - T_o)\delta}{\lambda_s(T_o - T_w)}\right\} \\ &- \frac{2i\rho_s + \rho_s c_s(T_o - T_w)}{2\alpha_o(T_f - T_o)} \cdot \delta \end{aligned} \tag{2.6}$$

This equation is not quadratic. Usually, the melts to be solidified have temperatures that exceed the melting point by 5, 10, or 15 K. This means in practice that the solidification times predicted by Eqs. (2.3), (2.5), and (2.6) are not widely different. This will be shown via a worked example.

Example 1

An organic melt is solidified according to the Cases 1, 2, and 3. As for Cases 1 and 2, the melt is poured on a thin (e.g., 1 mm) aluminum plate cooled by flashing propylene (5°C). The heat transfer to the air is neglected. Regarding Case 3, the thin aluminum plate is immersed in the stirred melt while the layer grows from below. The upper side is cooled. In Cases 2 and 3, the melt is "superheated" by 10 K. $\alpha_o = 100\ \mathrm{W \cdot m^{-2} \cdot K^{-1}}$ for Case 3. The times to grow a 3-mm layer are calculated.

Physical properties

$T_o = 132°\mathrm{C}$
$i = 209{,}340\ \mathrm{J \cdot kg^{-1}}$
$\rho_s = 1200\ \mathrm{kg \cdot m^{-3}}$
$c_l = 2800\ \mathrm{J \cdot kg^{-1} \cdot K^{-1}}$
$c_s = 1900\ \mathrm{J \cdot kg^{-1} \cdot K^{-1}}$
$\lambda_s = 0.174\ \mathrm{W \cdot m^{-1} \cdot K^{-1}}$

Case 1

$$t = \left\{\frac{209,340 \cdot 1200}{2(132-5)0.174} + \frac{1200 \cdot 1900}{4 \cdot 0.174}\right\}0.003^2 = 80.6 \text{ sec}$$

Case 2

$$t = \left[\frac{\{209,340 + 2800(142-132)\}1200}{2(132-5)0.174} + \frac{1200 \cdot 1900}{4 \cdot 0.174}\right]0.003^2$$

$$= 87.5 \text{ sec}$$

Case 3

$$t = -\frac{0.174(132-5)\{2 \cdot 209,340 \cdot 1200 + 1200 \cdot 1900(132-5)\}}{2 \cdot 100^2(142-132)^2}$$

$$\times\ {}^{e}\log\left\{\frac{0.174(132-5) - 100(142-132)0.003}{0.174(132-5)}\right\}$$

$$-\frac{2 \cdot 209,340 \cdot 1200 + 1200 \cdot 1900(132-5)}{2 \cdot 100(142-132)} \cdot 0.003$$

$$= -8750.5 \cdot -0.1459 - 1188.0 = 1276.7 - 1188.0 = 88.7 \text{ sec}$$

Notes

- The results of the calculations make sense.
- Although Eq. (2.6) does not have a quadratic form the result is comparable with the results of the two quadratic equations.

2.5. SOLIDIFICATION ON A WALL WITH A VARIABLE TEMPERATURE

In this section, the resistances to heat transfer of the metal wall and the cooling water film are not neglected. The basic equation is Eq. (2.2):

$$\frac{\lambda_s(T_o - T_w)}{\delta} \cdot dt = \alpha_o(T_f - T_o)dt + i\rho_s \cdot d\delta + \frac{\rho_s c_s d\{(T_o - T_w)\delta\}}{2}$$

The meaning of this equation was explained in Section 2.3.

Again, three cases will be considered:

- Case 1 concerns the situation in which the melt is "superheated", well-mixed and in contact with the growing layer (e.g., a drum flaker that picks up material from a pan).
- Case 2 addresses the situation in which the melt is at its melting point and is poured on a metal surface (of, e.g., a belt cooler) while there is no heat loss to the air.
- Case 3 is comparable to Case 2, however, the melt is "superheated". The melt is stagnant and nonconductive.

Case 1

First, the resistances to heat transfer offered by the cooling agent and the metal are combined:

$$\alpha_k = \frac{1}{\dfrac{1}{\alpha_c} + \dfrac{\delta_w}{\lambda_w}}$$

Second, a set of dimensionless numbers is defined.

$$\chi = \frac{\alpha_k \delta}{\lambda_s} \text{ (dimensionless layer thickness)}$$

$$\mu = \frac{c_s(T_o - T_c)}{2i}$$

$$\theta = \frac{T_f - T_o}{T_o - T_c} \cdot \frac{\alpha_o}{\alpha_k}$$

$$\tau = \frac{\alpha_k^2(T_o - T_c)}{i\rho_s\lambda_s} \cdot t \text{ (dimensionless time)}$$

Note that the expression for χ contains the ratio of the resistances to heat transfer offered by the solid (crystalline) layer and the combination metal wall/cooling medium. The next equation is used:

$$\alpha_k(T_w - T_c) = \frac{\lambda_s(T_o - T_w)}{\delta} \tag{2.7}$$

Note that this equation does not take the enthalpy change of the metal wall into account. For belt coolers, this effect is small for two reasons:

- The belts are thin, e.g., 1 mm.
- The heat capacity of the belt material is small, e.g., 0.5 kJ $\cdot$ kg^{-1} $\cdot$ K^{-1}.

For drum flakers, the second aspect also applies; however, the first does not. When the drum dips into the melt, the outer drum wall temperature is approximately equal to the delivery temperature of the flakes while the inner drum wall temperature is approximately equal to the cooling water temperature (Griesser, 1973). When the drum contacts the melt, the outer drum wall temperature rises to the melting point of the melt. At this point in time, the T/δ graph for the metal wall will not be linear. The outer drum wall temperature will fall rapidly as the layer of solidified material develops. However, the drum material leaving the pan is warmer than the drum material entering the pan. Thus, heat from the melt is stored in the metal wall making the heat transfer through the metal wall and to the cooling water more effective than suggested by the heat transfer coefficient α_k. Thus, on using Eq. (2.7), a conservative approach is followed. The effect of neglecting the enthalpy change of the metal wall is more important for thin layers (e.g., 0.5 mm) than for relatively thick layers (1–3 mm). An analogy is the pouring of some liquid melt from a burning candle on the skin. Initially, a sensation of pain is felt; however, the feeling disappears quickly as a solid layer develops. The solid layer effectively insulates the skin from the liquid melt. Equation (2.7) can be differentiated to:

$$dT_w = -\frac{\alpha_k}{\lambda_s} \cdot \frac{(T_w - T_c)^2}{(T_o - T_c)} \cdot d\delta \tag{2.8}$$

Integration of this equation results in:

$$\frac{1}{T_w - T_c} - \frac{1}{T_o - T_c} = \frac{\alpha_k \delta}{\lambda_s(T_o - T_c)} \tag{2.9}$$

We will now focus on Eq. (2.2). On substituting Eq. (2.7) and the definition of τ into the left-hand side of Eq. (2.2), $[(T_w - T_c)i\rho_s\lambda_s/\{\alpha_k(T_o - T_c)\}]\, d\tau$

is obtained. Likewise, the first part of the right-hand side (RHS) of Eq. (2.2) becomes $[\alpha_o(T_f - T_o)i\rho_s\lambda_s/\{\alpha_k^2(T_o - T_c)\}]d\tau$ by using the definition of τ. The second part of the RHS can be written as $(i\rho_s\lambda_s/\alpha_k)d\chi$ by substituting $\chi = \alpha_k\delta/\lambda_s$. Finally, the third part of the RHS of Eq. (2.2) is equal to $\{\rho_s c_s\lambda_s/(2\alpha_k)\}[\{(T_w - T_c)^2/(T_o - T_c)\}\chi + T_o - T_w]d\chi$. This expression is obtained by differentiating, substituting Eq. (2.8) and using the definition of χ. The result is:

$$\frac{(T_w - T_c)i\rho_s\lambda_s}{\alpha_k(T_o - T_c)} \cdot d\tau = \frac{\alpha_o(T_f - T_o)i\rho_s\lambda_s}{\alpha_k^2(T_o - T_c)} \cdot d\tau + \frac{i\rho_s\lambda_s}{\alpha_k} \cdot d\chi + \frac{\rho_s c_s\lambda_s}{2\alpha_k} \times \left\{\frac{(T_w - T_c)^2}{T_o - T_c} \cdot \chi + T_o - T_w\right\}d\chi \tag{2.10}$$

The combination of Eq. (2.7) and the definition of χ results in $\chi = \frac{T_o - T_w}{T_w - T_c}$. This simple relationship can be transformed into four equally simple relationships:

$$\frac{1}{\chi} = \frac{T_w - T_c}{T_o - T_w}, \quad 1 + \chi = \frac{T_o - T_c}{T_w - T_c}, \quad \frac{1}{1+\chi} = \frac{T_o - T_w}{T_w - T_c}, \quad \text{and}$$

$$\frac{1}{\chi} + 1 = \frac{T_o - T_c}{T_o - T_w}$$

Now, multiplying Eq. (2.10) by $\alpha_k/(i\rho_s\lambda_s)$, using the five simple relations, substituting the definition of θ into the second part, and the definition of μ into the third part of the RHS of Eq. (2.10) results in:

$$d\tau = \left\{\frac{\chi(\mu + 1) + 1}{1 - \theta(\chi + 1)} + \frac{\mu\chi}{\chi + 1 - \theta(\chi + 1)^2}\right\}d\chi \tag{2.11}$$

Integration of Eq. (2.11) gives:

$$\int d\tau = \int \frac{\mu\chi \cdot d\chi}{1 - \theta(\chi + 1)} + \int \frac{(\chi + 1)d\chi}{1 - \theta(\chi + 1)} + \int \frac{\mu\chi \cdot d\chi}{\chi + 1 - \theta(\chi + 1)^2} \tag{2.12}$$

The limits are: $\chi = \chi$ when $\tau = \tau$, and $\chi = 0$ when $\tau = 0$. The first part of the RHS of Eq. (2.12) can be solved by putting $\chi + 1 = y$. Thus, $d\chi = dy$ and $\chi = y - 1$. $\int \mu\chi \cdot d\chi/\{1 - \theta(\chi + 1)\} = \mu \int (y - 1) \cdot dy/(1 - \theta y) = \mu \int y \cdot dy/(1 - \theta y) - \mu \int dy/(1 - \theta y)$. The latter integral can be solved easily (integration between $y = 1$ ($\chi = 0$) and $y = \chi + 1$). The outcome is $(\mu/\theta)\ {}^e\log\{1 - \theta(\chi + 1)\}/(1 - \theta)$. The other integral can be solved by putting $1 - \theta y = z$. Thus, $y = (1 - z)/\theta$

and $dy = -dz/\theta$. The integration occurs between $z = 1 - \theta\,(y = 1)$ and $z = 1 - \theta \cdot (\chi + 1)$. The outcome is:

$$-\frac{\mu}{\theta^2}\left\{{}^{e}\log\left(1 - \frac{\theta\chi}{1-\theta}\right) + \theta\chi\right\}$$

Addition results in: $-(\mu/\theta^2 - \mu/\theta)\ {}^{e}\log\{1 - \theta(\chi + 1)\}/(1 - \theta) - \mu\chi/\theta$.
The second part of the RHS of Eq. (2.12) can be integrated as follows:

$$\int \frac{(\chi + 1)d\chi}{1 - \theta(\chi + 1)} = \int \frac{\chi d\chi}{1 - \theta(\chi + 1)} + \int \frac{d\chi}{1 - \theta(\chi + 1)}$$

$$y = 1 - \theta(\chi + 1),\ d\chi = -\frac{dy}{\theta},\ \text{and}\ \chi = -\frac{y}{\theta} - 1 + \frac{1}{\theta}$$

Substitution results in:

$$\int \frac{dy}{\theta^2} + \int \frac{dy}{\theta y} - \int \frac{dy}{\theta^2 y} - \int \frac{dy}{\theta y} = -\frac{\chi}{\theta} - \frac{1}{\theta^2}\,{}^{e}\log\left(1 - \frac{\theta\chi}{1 - \theta}\right)$$

Integration occurred between $y = 1 - \theta(\chi + 1)$ and $y = 1 - \theta$.
Finally, the third part of the RHS of Eq. (2.12) is solved.

$$\mu \int \frac{\chi \cdot d\chi}{\chi + 1 - \theta(\chi + 1)^2} = \mu \int \frac{(y - 1)dy}{y - \theta y^2} \quad \text{if } \chi = y - 1$$

$$\mu \int \frac{(y - 1)dy}{y(1 - \theta y)} = \mu \int \frac{dy}{1 - \theta y} - \mu \int \frac{dy}{y(1 - \theta y)}$$

The first integral can be integrated to $(-\mu/\theta)\ {}^{e}\log\{1 - \theta\chi/(1 - \theta)\}$ (integration between $y = 1$ and $y = \chi + 1$).

The second integral can be solved as follows:

$$\frac{1}{y(1 - \theta y)} = \frac{A}{y} + \frac{B}{1 - \theta y} \quad \text{(expression in partial fractions)}$$

$A = 1$ and $B = \theta$

Thus, $1/\{y(1 - \theta y)\} = 1/y + \theta/(1 - \theta y)$

$$-\mu \int \frac{dy}{y(1 - \theta y)} = -\mu \int \frac{dy}{y} - \mu \int \frac{\theta dy}{(1 - \theta y)}$$

The outcome is: $-\mu \cdot {}^{e}\log(\chi + 1) + \mu \cdot {}^{e}\log\left(1 - \frac{\theta\chi}{1 - \theta}\right)$ (integration between $y = 1$ and $y = \chi + 1$).

Addition: $-\mu \cdot {}^{e}\log(\chi + 1) - \mu\left(\frac{1}{\theta} - 1\right){}^{e}\log\left(1 - \frac{\theta\chi}{1 - \theta}\right)$

The integration of Eq. (2.12) results in:

$$\tau = -\frac{1}{\theta^2}\ {}^{e}\log\left(1-\frac{\theta\chi}{1-\theta}\right)-\frac{\chi}{\theta}$$
$$+\mu\left\{-\left(\frac{1}{\theta^2}-1\right)\ {}^{e}\log\left(1-\frac{\theta\chi}{1-\theta}\right)-\frac{\chi}{\theta}-\ {}^{e}\log(\chi+1)\right\}. \qquad (2.13)$$

This is the complete equation for drum flakers. However, it should be remembered that linear temperature profiles have been assumed. The third part of the RHS (the part multiplied with μ) concerns the enthalpy change of the solid layer. Gregorig (1976) uses the definitions:

$$F_0 = -\frac{1}{\theta^2}\ {}^{e}\log\left(1-\frac{\theta\chi}{1-\theta}\right)-\frac{\chi}{\theta}$$

$$F_1 = -\left(\frac{1}{\theta^2}-1\right)\ {}^{e}\log\left(1-\frac{\theta\chi}{1-\theta}\right)-\frac{\chi}{\theta}-\ {}^{e}\log(\chi+1)$$

$$\tau = F_0+\mu F_1$$

It is possible to make an estimate of the consequences when the enthalpy change of the solid layer is neglected. For very long times, the ratio $(F_0+\mu F_1)/F_0$ becomes

$$\frac{-\dfrac{1}{\theta^2}-\mu\left(\dfrac{1}{\theta^2}-1\right)}{-\dfrac{1}{\theta^2}} = 1+\mu(1-\theta^2).$$

This is explained as follows. Very long times can be calculated from Eq. (2.13) only if $1-\frac{\theta\chi}{1-\theta}$ approaches zero. $-{}^{e}\log\left(1-\frac{\theta\chi}{1-\theta}\right)$ then becomes very large. In comparison to the parts containing this logarithmic expression, the other parts of the RHS of Eq. (2.13) can be neglected.

Example 2

The effect of the enthalpy change (specific heat) is calculated for a situation that is analogous to the third situation of Example 1 in Sec. 2.4. The metal wall temperature is no longer constant but decreases as time proceeds.

Additional data

$\alpha_c = 1000\ \mathrm{W\cdot m^{-2}\cdot K^{-1}}$ (Preger, 1969, Maschinenmarkt)
$\delta_w = 0.01$ m
$\lambda_w = 40\ \mathrm{W\cdot m^{-1}\cdot K^{-1}}$

The effect of the enthalpy change is calculated for very long times.

$$\frac{1}{\alpha_k}=\frac{1}{\alpha_c}+\frac{\delta_w}{\alpha_w},\quad \frac{1}{\alpha_k}=\frac{1}{1000}+\frac{0.01}{40}=0.00125$$

$$\alpha_k = 800\ \text{W}\cdot\text{m}^{-2}\cdot\text{K}^{-1}$$

$$\theta = \frac{T_f - T_o}{T_o - T_c}\cdot\frac{\alpha_o}{\alpha_k} = \frac{142-132}{132-5}\cdot\frac{100}{800} = 0.00984$$

$$\mu = \frac{c_s(T_o - T_c)}{2i} = \frac{1900(132-5)}{2\cdot 209,340} = 0.576$$

$$1+\mu(1-\theta^2) = 1 + 0.576(1-0.00984^2) = 1.576$$

Neglecting the specific heat leads to a crystallization time that is 1.576 times too short. Clearly, the enthalpy change should not be neglected.

Again, inspection of Eq. (2.13) shows that a very large τ can be obtained only if $1-\theta\chi/(1-\theta)$ approaches zero. The physical meaning is that the growth of the layer stops when the layer has reached a certain thickness. At that point, the heat flow from the "superheated" melt due to convective heat transfer can just be carried away by the solidified layer. This thickness is called the limiting layer thickness (χ_m).

$$1 - \frac{\theta\chi_m}{1-\theta} = 0 \quad \text{leads to:}$$

$$\chi_m = \frac{1}{\theta} - 1 \tag{2.14}$$

Example 3

It is required to calculate the limiting layer thickness for the previous example in this section. $\theta = 0.00984$

$$\chi_m = \frac{1}{\theta} - 1 = 100.63,\quad \chi_m = \frac{\alpha_k\delta_m}{\lambda_s}$$

$$\delta_m = \frac{\chi_m\lambda_s}{\alpha_k} = \frac{100.63\cdot 0.174}{800} = 0.022\ \text{m}\ (22\ \text{mm})$$

The concept of the limiting layer thickness will be discussed further in Sec. 2.6.

Example 4

An organic melt is solidified via a drum flaker.

$T_f = 142°\text{C}$
$T_c = 15°\text{C}$
$\alpha_o = 100\ \text{W}\cdot\text{m}^{-2}\cdot\text{K}^{-1}$
$\alpha_c = 1000\ \text{W}\cdot\text{m}^{-2}\cdot\text{K}^{-1}$
$\delta_w = 0.01\ \text{m}$
$\lambda_w = 40\ \text{W}\cdot\text{m}^{-1}\cdot\text{K}^{-1}$

Physical properties

$T_o = 132°C$
$i = 209{,}340\ J \cdot kg^{-1}$
$\rho_s = 1200\ kg \cdot m^{-3}$
$c_l = 2800\ J \cdot g^{-1} \cdot K^{-1}$
$c_s = 1900\ J \cdot kg^{-1} \cdot K^{-1}$
$\lambda_s = 0.174\ W \cdot m^{-1} \cdot K^{-1}$

The time to grow a 1-mm layer is calculated.

$$\frac{1}{\alpha_k} = \frac{1}{\alpha_c} + \frac{\delta_w}{\lambda_w};\ \frac{1}{\alpha_k} = \frac{1}{1000} + \frac{0.01}{40}$$

$$\alpha_k = 800\ W \cdot m^{-2} \cdot K^{-1}$$

$$\theta = \frac{T_f - T_o}{T_o - T_c} \cdot \frac{\alpha_o}{\alpha_k} = \frac{142 - 132}{132 - 15} \cdot \frac{100}{800} = 0.0107$$

$$\chi = \frac{\alpha_k \cdot \delta}{\lambda_s} = \frac{800 \cdot 0.001}{0.174} = 4.598$$

$$\mu = \frac{c_s(T_o - T_c)}{2i} = \frac{1900(132 - 15)}{2 \cdot 209,340} = 0.5310$$

$$F_0 = -\frac{1}{\theta^2}\ {}^e\log\left(1 - \frac{\theta\chi}{1-\theta}\right) - \frac{\chi}{\theta} = 15.82$$

$$F_1 = -\left(\frac{1}{\theta^2} - 1\right)\ {}^e\log\left(1 - \frac{\theta\chi}{1-\theta}\right) - \frac{\chi}{\theta} - {}^e\log(\chi + 1) = 14.05$$

$$\tau = F_0 + \mu F_1 = 23.28$$

$$t = \frac{\tau i \rho_s \lambda_s}{\alpha_k^2(T_o - T_c)} = \frac{23.28 \cdot 209,340 \cdot 1200 \cdot 0.174}{800^2(132 - 15)} = 13.6\ \text{sec}$$

The wall temperature T_w is 36°C after 13.6 sec.

Equation (2.13): $\dfrac{1}{T_w - T_c} - \dfrac{1}{T_o - T_c} = \dfrac{\alpha_k \delta}{\lambda_s(T_o - T_c)}$;

$$\frac{1}{36 - 15} - \frac{1}{132 - 15} = \frac{800 \cdot 0.001}{0.174(132 - 15)}$$

Notes

(1) For a 2-mm layer, a solidification time of 49.3 sec and a wall temperature T_w of 27°C can be calculated. It appears that the "quadratic growth law" (see Sec. 2.3) is approximately applicable.

(2) If the solidification angle of the drum flaker is 81.5°, then the rotational speed is 1 rpm.

(3) In Example 4, the entrainment of liquid from the pan was neglected. This aspect will be discussed in Chapter 4.

(4) The heat transfer coefficient from the melt to the solid layer was arbitrarily fixed at 100 $\text{W} \cdot \text{m}^{-2} \cdot \text{K}^{-1}$. It will be shown in Chapter 4 how the heat transfer coefficient can be estimated.

Case 2

Equation (2.12) becomes ($\theta = 0$):

$$\int d\tau = \int \mu\chi \cdot d\chi + \int (\chi + 1) d\chi + \int \frac{\mu\chi \cdot d\chi}{(\chi + 1)} \tag{2.15}$$

First part of the RHS: $\mu\chi^2/2$.

Second part of the RHS: $\chi + \chi^2/2$.

The third part of the integral can be solved by putting $y = \chi + 1$ ($\chi = y - 1$, $dy = d\chi$).

Result: $\mu\chi - \mu \cdot {}^e\log(\chi + 1)$

$$\tau = \chi + \frac{\chi^2}{2} + \mu\left\{\chi + \frac{\chi^2}{2} - {}^e\log(\chi + 1)\right\} \tag{2.16}$$

Example 5

See Example 4. The time to grow a 1-mm layer is calculated. The melt temperature is 132°C.

$\chi = 4.598$
$\mu = 0.5310$

$$\tau = \chi + \frac{\chi^2}{2} + \mu\left\{\chi + \frac{\chi^2}{2} - {}^e\log(\chi + 1)\right\} = 15.17 + 0.5310 \cdot 13.45$$

$$= 22.3$$

$$t = \frac{\tau i \rho_s \lambda_s}{\alpha_k^2 (T_o - T_c)} = 13.0 \text{ sec}$$

Notes

(1) For a 2-mm layer, a solidification time of 45.2 sec can be calculated. Again, the "quadratic growth law" is approximately applicable.

(2) The solidification time is shorter than the solidification time calculated in Example 4. This makes sense as the melt is not "superheated".

Case 3

Equation (2.16) can be applied. Use these definitions:

$$\tau' = \frac{\alpha_k^2 (T_o - T_c)}{\{i + c_l(T_f - T_o)\}\rho_s \lambda_s} \cdot t$$

$$\mu' = \frac{c_s(T_o - T_c)}{2\{i + c_l(T_f - T_o)\}}$$

$$\tau' = \chi + \frac{\chi^2}{2} + \mu' \left\{ \chi + \frac{\chi^2}{2} - {}^e\log(\chi + 1) \right\} \tag{2.17}$$

In this approach, the sensible melt heat is added to the heat of fusion.

Example 6

See Examples 4 and 5. The time to grow a 1-mm layer is calculated. The melt temperature is 142°C.

$$\chi = 4.598$$

$$\mu' = \frac{1,900(132 - 15)}{2\{209,340 + 1,900(142 - 132)\}} = 0.487$$

$$\tau' = 21.72$$

$$t = 14.4 \text{ sec}$$

Concluding Remarks

(1) The solidification times calculated in this section are not very different. It takes 13.6 sec to grow a 1-mm layer from a well-mixed "superheated" melt. It takes 13.0 sec to obtain a 1-mm layer from a melt at its melting point. The small difference can be explained by the "superheat" of 10 K only. Furthermore, 14.4 sec is calculated for a 1-mm layer from a stagnant "superheated" melt.

(2) The "quadratic growth law" is approximately applicable (see Sec. 2.3). Empirically, it was found for drum flakers that the layer thickness is approximately inversely proportional to the root of the rotational speed:

$$\delta = An^{-0.5}$$

It follows:

$$\delta = Bt^{0.5}$$

(Preger, 1968, 1970, 1980).

Preger reports his results for sulfur, sodium sulfide (60%), caustic soda (98%), a synthetic resin, and two organic materials.

Deviations can be explained as follows:

- The metal wall temperature is not constant during the solidification.
- The rotating drum entrains melt from the pan. This effect is more pronounced for high drum circumferential velocities than for small velocities (e.g., $<0.1\ \mathrm{m \cdot sec^{-1}}$). Viscous melts are more readily entrained than thin melts.
- α_o increases when the rotational speed increases, and this causes the layer thickness to become relatively small at high rotational speeds. However, this effect is small when the usual "superheats" prevail, i.e., 5, 10, or 15 K.

2.6. THE CONCEPT OF THE LIMITING LAYER THICKNESS

Case 1 of Sec. 2.5 concerns the situation in which a layer grows on a metal surface while the growing layer is in contact with a well mixed melt. The resistance to heat transfer of the deposited layer increases as the layer thickness increases. Eventually, a situation appears that the growth stops. It is then possible to write (Fig. 2.1):

$$\alpha_o(T_f - T_o) = \frac{\lambda_s}{\delta_m}(T_o - T_w) = \frac{\lambda_w}{\delta_w}(T_w - T_i) = \alpha_c(T_i - T_c) \tag{2.18}$$

The process of heat transfer is now stationary.

Note that the heat resistances of the metal wall and the boundary layer of the cooling water are not neglected. Equation (2.18) can be restated as follows:

$$\alpha_o(T_f - T_o) = \frac{\lambda_s}{\delta_m}(T_o - T_w) \tag{2.19}$$

$$\alpha_o(T_f - T_o) = \frac{\lambda_w}{\delta_w}(T_w - T_i) \tag{2.20}$$

$$\alpha_o(T_f - T_o) = \alpha_c(T_i - T_c) \tag{2.21}$$

Equation (2.19) can be rewritten as $T_w = \dfrac{-\alpha_o\delta_m(T_f - T_o)}{\lambda_s} + T_o$.

Equation (2.21) can be rewritten as $T_i = \dfrac{\alpha_o(T_f - T_o)}{\alpha_c} + T_c$.

Substitution of these two expressions into Eq. (2.20) results in:

$$\delta_m = \frac{\lambda_s(T_o - T_c)}{\alpha_o(T_f - T_o)} - \lambda_s\left(\frac{1}{\alpha_c} + \frac{\delta_w}{\lambda_w}\right) \tag{2.22}$$

Example 7

The limiting layer thickness for the solidification described in Example 2 of this chapter is calculated by using Eq. (2.22).

$$\delta_m = \frac{0.174(132 - 5)}{100(142 - 132)} - 0.174\left(\frac{1}{1000} + \frac{0.01}{40}\right) = 0.022 \text{ m (22 mm)}$$

As soon as the layer thickness has grown to 22 mm, the melt remelts any material further solidified. Note that the result equals the result calculated in Example 3 of this chapter. The process conditions and the physical data are equal in both cases.

The concept of the limiting layer thickness can also be applied to Case 3 of the solidifications described in Sec. 2.4. Using the data in Example 1 of this chapter, $\delta_m = 0.022$ m is calculated. In this case, the limiting layer thickness is marginally greater than in Example 7.

2.7. COOLING ON A WALL WITH A CONSTANT TEMPERATURE

The theory discussed in this section is applicable for the cooling of slabs on belt coolers that are cooled efficiently by vigorous water sprays. The theory can, generally speaking, not be used for drum flakers. The reason is that drum flakers have rather thick walls (10–20 mm) while the belt of a belt cooler has a thickness of, e.g., 1 mm. Thus, belts quickly loose their sensible heat as they are thin and hence have a small heat capacity. Flaker walls do not cool down so readily. The cooling of slabs on drum flakers will be discussed in Chapter 4.

It is important to distinguish between two temperatures of the material when it leaves a cooler:

- The temperature of the uncooled surface
- The average temperature.

The temperature of the uncooled surface is important because it determines the caking properties. The average temperature should be known because it is needed for the heat balance. The former temperature is more important than the latter one for the design of belt coolers. Therefore, the approach to arrive at the temperature of the uncooled surface is discussed first. The cooling of a solidified layer will be considered as the unidirectional flow of heat through a body with plane parallel faces a distance h apart. The heat flow is normal to these faces, and the temperature of the body is initially constant throughout.

It can be assumed that the temperature of the body equals the melting point initially. This is a conservative approach. The assumption of an initial uniform temperature being equal to the arithmetic average of the cooling water temperature and the melting point is a less conservative approach. However, the temperature scale will be so chosen that this uniform initial temperature is zero. Furthermore, it is assumed that at time $t=0$, one face (at $\delta=0$) is brought into contact with a source at a constant temperature T_W, and the other face (at $\delta=h$) will be assumed to be perfectly thermally insulated.

The basic differential equation for the process is Eq. (2.1):

$$\frac{\partial T}{\partial t} = a\frac{\partial^2 T}{\partial \delta^2}.$$

The boundary conditions are:

$$\begin{cases} t=0, & (T-T_o)=0 \\ t>0, & T=(T_W-T_o) \text{ when } \delta=0 \\ t>0, & \partial T/\partial\delta = 0 \text{ when } \delta = h. \end{cases}$$

The temperature T of the distant (insulated) face is given by

$$(T-T_o) = \sum_{N=0}^{N=\infty}(-1)^N \cdot (T_W - T_o)2 \cdot \operatorname{erfc}\frac{(2N+1)h}{2\sqrt{at}} \tag{2.23}$$

(Coulson and Richardson, 1999).

$$\operatorname{erfc} z = \frac{2}{\sqrt{\pi}}\int_z^{\infty} e^{-\xi^2}\cdot d\xi$$

Coulson and Richardson (1999) give values erf $z=1-\operatorname{erfc} z$. Table 2.1 contains some values of erfc z. In many instances, the second and higher terms are negligible compared with the first term. Whether this is true depends on t and can be established by inspection.

The average temperature is represented by $\bar{T}$.

Likewise, a relation can be arrived at by integrating Eq. (2.1). The following relation is proposed:

$$\frac{T_W-\bar{T}}{T_W-T_o} = \frac{8}{\pi^2}\left(e^{-bFo} + \frac{1}{9}e^{-9bFo} + \frac{1}{25}e^{-25bFo} + \cdots\right) \tag{2.24}$$

$$Fo = \frac{at}{h^2}$$

$$b = \left(\frac{\pi}{2}\right)^2 = 2.467$$

(McCabe et al., 2001)

Equation (2.24) is applicable for the cooling of an infinite slab of thickness $2h$ from both sides by a medium at constant surface temperature.

TABLE 2.1 erfc z

z	erfc z
0	1.000
0.1	0.888
0.2	0.777
0.3	0.671
0.4	0.572
0.5	0.480
0.6	0.396
0.7	0.322
0.8	0.258
0.9	0.203
1.0	0.157
1.1	0.120
1.2	0.090
1.3	0.066
1.4	0.048
1.5	0.034
1.6	0.024
1.7	0.016
1.8	0.011
1.9	0.007
2.0	0.005
∞	0.000

This equation is also valid for the cooling of an infinite slab of thickness h from one side, while the other side is thermally perfectly insulated. The reason is that the heat flux through the central plane of a slab of thickness $2h$, which is cooled from both sides, is zero. When Fo is greater than about 0.1, only the first term of the series in Eq. (2.24) is significant and the other terms can be ignored. The time required to change the temperature from T_o to $\overline{T}$ can then be found by rearranging Eq. (2.24). All except the first term are omitted.

$$t = \frac{1}{a}\left(\frac{2h}{\pi}\right)^2 {}^{e}\log\frac{8(T_W - T_o)}{\pi^2(T_W - \overline{T})} \tag{2.25}$$

Example 8

Reference is made to Example 1 in Section 2.4. However, the material is cooled on a belt cooler. Cooling water at 15°C is used. Equation (2.23) is used to calculate the cooling time to obtain a surface temperature of 50°C.

$T_W = 15°C$

$$h = 0.003 \text{ m}$$

$$a = \frac{\lambda_s}{c_s \rho_s} = \frac{0.174}{1900 \cdot 1200} = 7.63 \cdot 10^{-8} \text{ m}^2 \cdot \text{sec}^{-1}$$

An approximate answer is obtained by taking the first term only.

$$(50 - 132) = (15 - 132)2 \cdot \text{erfc}\frac{0.003}{2\sqrt{7.63 \cdot 10^{-8} \cdot t}}$$

$$0.350 = \text{erfc}\,(5.43 \cdot t^{-0.5})$$
$$t = 68 \text{ sec}$$

It will be checked whether the second term is small compared to the first term.

$$\text{First Term}: \quad (15 - 132)2 \cdot \text{erfc}\frac{0.003}{2\sqrt{7.63 \cdot 10^{-8} \cdot 68}} = -82$$

$$\text{Second Term}: -(15 - 132)2 \cdot \text{erfc}\frac{3 \cdot 0.003}{2\sqrt{7.63 \cdot 10^{-8} \cdot 68}} = 1.2$$

It is therefore allowed to take only the first term into consideration.

Furthermore, Eq. (2.25) is used for the calculation of the average temperature.

$$Fo = \frac{7.63 \cdot 10^{-8} \cdot 68}{(0.003)^2} = 0.576$$

Equation (2.25) can be applied because Fo is greater than approximately 0.1.

$$68 = \frac{1}{7.63 \cdot 10^{-8}} \left(\frac{2 \cdot 0.003}{\pi}\right)^2 {}^{e}\log\frac{8(15 - 132)}{\pi^2(15 - \overline{T})}$$

$$1.422 = {}^{e}\log - \frac{94.837}{15 - \overline{T}}$$

$$\overline{T} = 38°\text{C}$$

Note

38 is not the arithmetic average of 15 and 50. The reason is that the uncooled surface is perfectly insulated:

$$\frac{\partial T}{\partial \delta} = 0 \quad \text{when} \quad \delta = h$$

LIST OF SYMBOLS

A	Proportionality constant in Eq. (2.4)
	Quantity on integrating by partial fractions (see Section 2.5)
a	Acceleration $[\mathrm{m} \cdot \mathrm{sec}^{-2}]$
	Thermal diffusivity $\{\lambda_s/(c_s\rho_s)\}$ $[\mathrm{m}^2 \cdot \mathrm{sec}^{-1}]$
B	Proportionality constant {see Eq. (2.4)}
	Quantity on integrating by partial fractions (see Section 2.5)
b	$(\pi/2)^2 = 2.467$
C_1	Auxiliary variable (see Sec. 2.4) $[\mathrm{m}^{-1}]$
C_2	Auxiliary variable (see Sec. 2.4) $[\mathrm{s} \cdot \mathrm{m}^{-2}]$
c_1	Melt specific heat $[\mathrm{J} \cdot \mathrm{kg}^{-1} \cdot \mathrm{K}^{-1}]$
c_s	Solid specific heat $[\mathrm{J} \cdot \mathrm{kg}^{-1} \cdot \mathrm{K}^{-1}]$
erf	Error function
erfc	1−erf
F_0	Gregorig function (see Sec. 2.5)
F_1	Gregorig function (see Sec. 2.5)
Fo	Fourier number (at/h^2)
h	Thickness of a cooled layer [m]
i	Heat of fusion $[\mathrm{J} \cdot \mathrm{kg}^{-1}]$
N	Serial number in Eq. (2.23)
n	Drum flaker rotational speed $[\mathrm{sec}^{-1}]$
s	Path traveled [m]
T	Temperature [°C]
ΔT	Temperature difference [K]
$\bar{T}$	Average slab temperature [°C]
T_c	Cooling medium temperature [°C]
T_f	Melt temperature [°C]
T_i	Metal wall temperature (medium-side) [°C]
T_o	Melting point [°C]
T_w	Metal wall temperature (process-side) [°C]
t	Time [sec]
	Time for one rotation [sec]
y	Auxiliary variable on integrating (see Sec. 2.5)
z	Auxiliary variable on integrating (see Secs. 2.4 and 2.5)
	Independent variable in the error function
	Part of the exponent in Eq. (2.4)
α_c	Heat transfer coefficient (medium-side) $[\mathrm{W} \cdot \mathrm{m}^{-2} \cdot \mathrm{K}^{-1}]$
α_k	Heat transfer coefficient (medium-side, includes metal wall resistance) $[\mathrm{W} \cdot \mathrm{m}^{-2} \cdot \mathrm{K}^{-1}]$
α_o	Heat transfer coefficient (process-side) $[\mathrm{W} \cdot \mathrm{m}^{-2} \cdot \mathrm{K}^{-1}]$
δ	Product layer thickness [m or mm]

δ_m	Limiting product layer thickness [m]
δ_w	Metal wall thickness [m]
θ	Dimensionless number
λ_s	Solid thermal conductivity $[W \cdot m^{-1} \cdot K^{-1}]$
λ_w	Metal wall thermal conductivity $[W \cdot m^{-1} \cdot K^{-1}]$
μ	Dimensionless number ("sensible heat/latent heat")
μ'	Dimensionless number ("sensible heat/latent heat" for "superheated" melt)
ξ	Auxiliary variable in the error function
ρ_s	Solid specific mass $[kg \cdot m^{-3}]$
τ	Dimensionless number ("time")
τ'	Dimensionless number ("time" for "superheated" melt)
χ	Dimensionless number ("layer thickness")
χ_m	Dimensionless limiting layer thickness

REFERENCES

Coulson, J. M. and Richardson, J. F. with Backhurst, J. R. and Harker, J. H. (1999). *Chemical Engineering*, Vol. 1. Oxford, England: Butterworth-Heinemann.

Gel'perin, N. I., Nosov, G. A., Parokonnyi, V. D. (1978). Crystallization of a superheated melt on cooled surfaces. *International Chemical Engineering* 18:129.

Gregorig, R. (1976). *Heat Exchangers*. Frankfurt am Main, Germany: Sauerländer. In German.

Griesser, L. (1973). Contribution to the calculation of drum flakers. *Aufbereitungs-Technik* 14:561. In German.

Heertjes, P. M., Ong Tjing Gie (1960). Crystallisation of water by unidirectional cooling. *British Chemical Engineering* 5:413.

McCabe, W. L., Smith, J. C., Harriott, P. (2001). *Unit Operations of Chemical Engineering*. New York: McGraw-Hill.

Preger, M. (1968). Behaviour of cooling cylinders on changing the operating conditions. *Aufbereitungs-Technik* 9:123. In German.

Preger, M. (1969). Operation and design of drum flakers. *Aufbereitungs-Technik* 10:39. In German.

Preger, M. (1969). Drum flakers as heat exchangers. *Maschinenmarkt* 75:1622. In German.

Preger, M. (1970). Application and designs of the drum flaker. *Aufbereitungs-Technik* 11:551. In German.

Preger, M. (1980). Drum flaker systems: status of the development and trends. *Maschinenmarkt* 86:80. In German.

3

Cooling Belts

3.1. INTRODUCTION

The theory for the design of belt coolers was discussed in Chapter 2. Sec. 3.2 contains a general description of belt coolers. Attention is paid to feeding devices, the belt itself, and discharge devices. It is possible to carry out tests on benchscale, and this item is discussed in Sec. 3.3. Three typical applications are dealt with in Sec. 3.4. Pastillation is gaining importance and this topic is reviewed in Sec. 3.5. Miscellaneous aspects of belt coolers (e.g., typical operational data) receive attention in Sec. 3.6.

The design method for belt coolers is summarized in Sec. 3.7 whereas Sec. 3.8 contains a verification of this model. The design of a typical cooling belt for the production of flakes is given in Sec. 3.9. Finally, Secs. 3.10 and 3.11 contain worked examples concerning the production of pastilles.

3.2. GENERAL DESCRIPTION

3.2.1. Introduction

A cooling belt is a moving, endless thin metal belt on which a melt can be solidified by spray cooling with water from below. A typical cooling belt system is depicted in Fig. 3.1. The feed tank contains the molten product. In many instances, the feed is blanketed with an inert gas to avoid degradation. The melt is pumped through a filter and to a rotary dropformer, a typical feeding device producing pastilles. The pastilles are deposited on the steel belt cooler, the heart of the system. When the flat belt surface is flexed to take the curvature of the terminal pulley, the shearing action between the solid pastille and the belt surface has the effect of causing the pastille to detach readily. Only a light plastic scraper is necessary if the pastilles do not fall away from the band before even reaching the scraper. Belts are quite thin (e.g., 1 mm thick), thus giving a very efficient heat transfer. They also adjust rapidly to a new temperature regime.

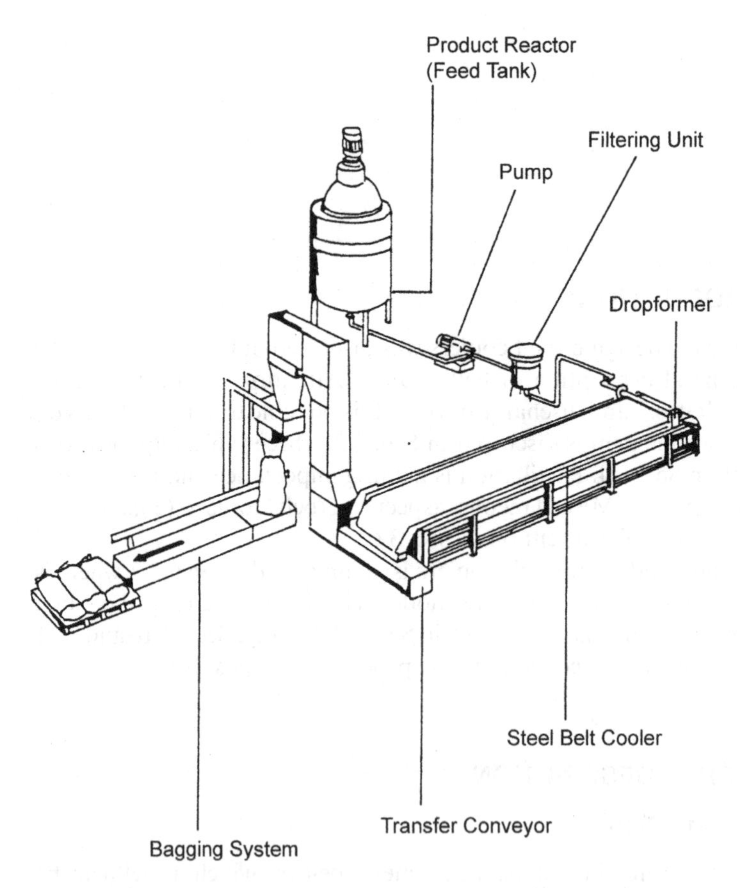

FIG. 3.1 A typical cooling belt system. (Courtesy of Sandvik Process Systems, Fellbach/Stuttgart, Germany.)

3.2.2. Feeding and Discharge Devices

Seven different feeding devices can be distinguished:

- Double roll feeder
- Overflow weir feeder
- Stripformer
- Rotary dropformer
- Plunger/nozzle
- Casting box feeder
- Extruder.

The first four feeding devices are depicted in Fig. 3.2.

The double roller is used for melts having viscosities up to 10^5 mPa · sec. The product is squeezed between two rollers to the required thickness and deposited on the steel belt. The rolls can be heated or cooled.

The overflow weir feeder can be used for viscosities up to 1000 mPa · sec. The sheet thickness is usually in the range of 1–3 mm. Retaining strips at the belt edges can be installed to prevent the melt from leaving the belt. It is also

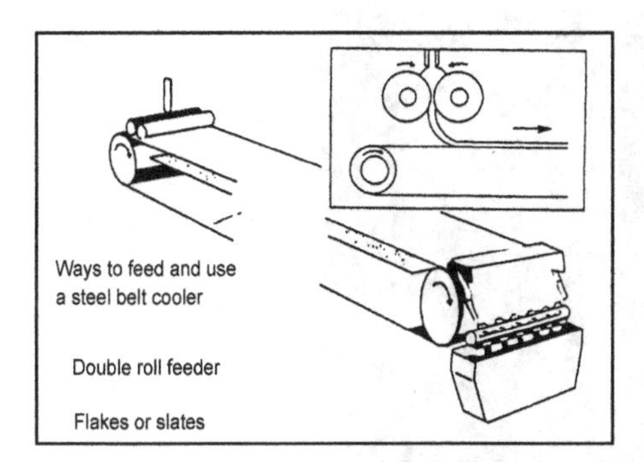

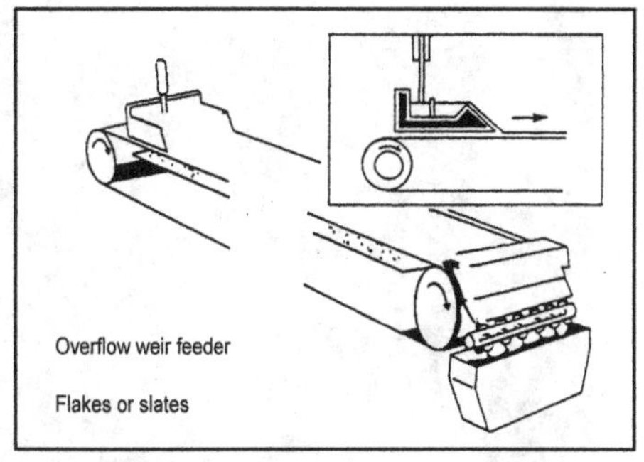

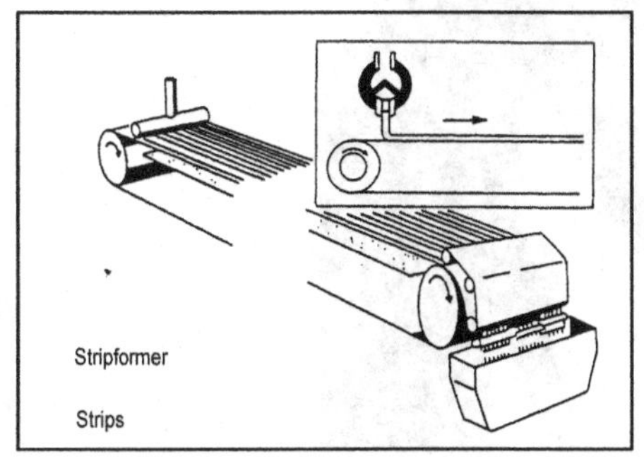

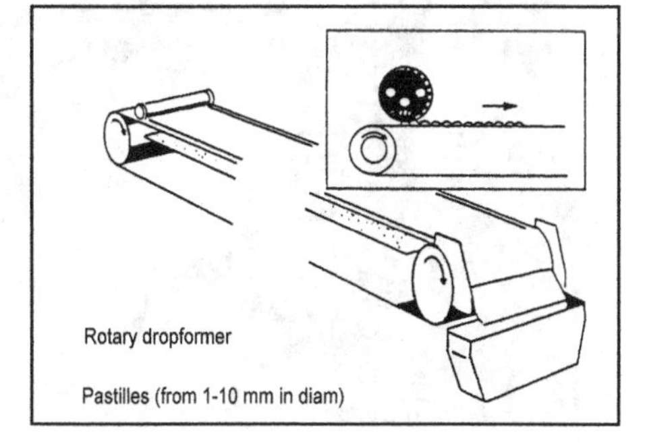

FIG. 3.2 Four different feeding devices. (Courtesy of Sandvik Process Systems, Fellbach/Stuttgart, Germany.)

possible to use air skirts to keep the product contained on the belt. Fig. 3.3 shows a solidified sheet produced by means of an overflow weir.

The heated stripformer is used for, for example, hot melts and atactic polypropylene. A typical purpose of making strips is facilitating subsequent particle size reduction. Thus, the installation of expensive granulators can often be avoided. Likewise, dust formation on processing brittle materials can be minimized.

The rotary dropformer deposits pastilles onto the belt. The diameter of the pastilles is in the range of 4–20 mm. Sandvik can also supply equipment to produce micropastilles having a diameter of 1 mm. Ideal pastilles have a hemispherical shape. The viscosity of the melt can be in the range of 10–20,000 mPa · sec. The feed temperature can range from, for example, 50 °C up to a maximum of 300 °C. Pastilles are often favored for these reasons:

- Practically dust-free
- Good flow characteristics
- Good caking properties (point contact rather than line or plane contacts).

FIG. 3.3 Sheet material produced on a belt cooler. (Courtesy of Sandvik Process Systems, Fellbach/Stuttgart, Germany.)

Rotary dropformers will be discussed further in Sec. 3.5. Eighty percent of the feeding devices sold by Sandvik are Rotoforms.

Two different plunger/nozzle systems are depicted in Fig. 3.4. Today, plungers and nozzles are largely replaced by rotary dropformers. The reason is that rotary dropforming equipment permits larger capacities per meter of belt width than plungers and nozzles. This will be discussed in Sec. 3.5. There is a niche where plungers and nozzles can still advantageously be applied. This niche is the pastillation of certain low-viscosity melts at low capacities. Here, the deposition of a droplet under gravity results in proper pastilles. Rotary dropformers can result in flatter structures because deposition is then accomplished under the combination of gravity and a small centrifugal force.

The Kaiser ZN system (Fig. 3.4) can be used for melts having a low viscosity (e.g., maleic acid anhydride and paraffin). The GS system is suitable for the pastillation of melts having an intermediate or high viscosity (hot-melts, resins). The simple plunger is now replaced by a plunger with a cylinder.

A casting box feeder is typically a box placed on the belt while the moving belt is the bottom of the feeder. The belt entrains a sheet of material from the box and the thickness can be adjusted by means of the casting bar. This feeding device can be used up to 40,000 mPa · sec.

An extruder can produce a sheet, slab, or rod. These extrudates can subsequently be cooled on a belt.

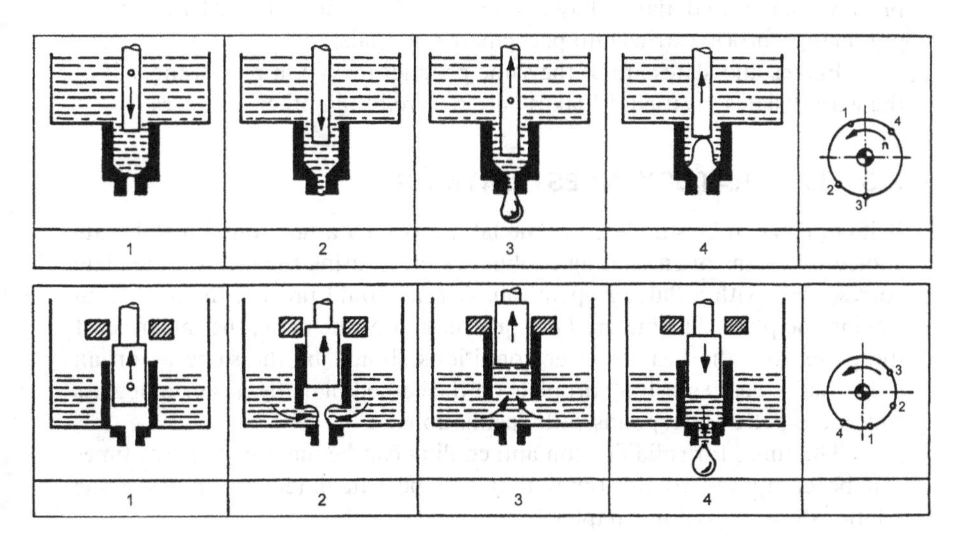

FIG. 3.4 Plungers and nozzles for feeding belt coolers. ZN for low-viscosity melts (upper part) and GS for medium viscosities (lower part). (Courtesy of Kaiser, Krefeld, Germany.)

The heat released during solidification and cooling is conducted through the belt and is transferred to cooling water sprayed against the underside. It may offer advantages for the crystallization process to have different cooling water temperatures in different zones (sections) of the belt. Usually, cooling water is used for cooling. The reason is that it is impractical to use other coolants because the system is not very closed on the medium side. (When solidifying on drum flakers, other coolants can be used.) Cooling water can be used on a once-through basis; recycling is also possible. Condensation of water from the air on the belt might occur if the cooling water temperature is lower than 15°C. This could contaminate the product. A countermeasure could be to enclose the belt and to condition the air coming in contact with the belt. Besides indirect cooling with cooling water, direct cooling by circulating air is possible (not very efficient). Some melts (e.g., maleic acid anhydride and naphthalene) tend to sublime and this gives rise to incrustations in the hood of the belt cooler.

Suction fans should be installed for melts generating a bad smell. Belt coolers can be equipped with gas-tight tunnels with an inert atmosphere.

The three different discharge devices are:

- Scrapers
- Breaker rolls
- Strip granulators.

The scrapers remove sheets and pastilles from the belt. Breaker rolls produce slates and flakes having sizes in the range of 1–20 mm. Strip granulators process strips into particulate materials.

Flakes and slates are of the same product form. On processing sulfur, the word "slates" is used. "Flakes" are referred to in other instances.

3.3. LABORATORY INVESTIGATIONS

Belt coolers can be simulated in the laboratory by a horizontal metal plate cooled by means of circulating cooling water. Pouring the melt on the plate corresponds with a flaking operation. A rim should prevent the melt from leaving the plate. The material of construction of the plate, the thickness of the plate, and the heat transfer conditions should be the same as on an industrial scale. Depositing droplets on the cooled plate by means of a glass rod or a pipette corresponds with pastillation.

The times for solidification and cooling can be measured. These times can be compared to the times that can be calculated by applying the relationships derived in Chapter 2.

The bonding forces between pastilles/flakes and metal surfaces can be measured.

The independent process variables that can be varied while checking their impact on the desired process results are:

- The feed temperature
- The temperature of the metal surface
- The application of a release agent (e.g., silicon oil) or coating with Teflon
- The partial precrystallization of the feed (see Chapter 6).

Many melts crystallize on the belt while other melts become amorphous solids. A criterion for crystallinity is the exhibition of a regular x-ray diffraction pattern. Some melts show the phenomenon of supercooling and this is discussed in Chapter 6. In principle, for crystalline materials, the cooling rate has an impact on product quality. Relatively slow cooling leads to relatively large crystals tending to be brittle. Some amorphous materials (e.g., certain resins) do not have a good belt contact if cooled too quickly.

Good flaking occurs when the melt flows out to give a layer of the desired thickness. The layer should solidify and cool uniformly while there is good belt contact. The adhesion should not be too good because it must be possible to remove the layer by means of a scraper without excessive dust generation. The application of a release agent is restricted to cases where contamination is not a problem. A well-known example is the application of a soap solution at the solidification of sulfur. A poor belt contact will give rise to long solidification times.

Good pastillation occurs if the pastilles have a hemispherical shape. Eight millimeters is a typical pastille diameter. The effect of the mentioned process variables on the pastillation process can be checked. The contact with the belt should neither be too good nor too bad. "Jumping" pastilles tend to loosen from the belt and to stay in touch via a point contact only. This will lead to long cooling times and low production intensities ($kg \cdot m^{-2} \cdot hr^{-1}$). Note that in stating production intensities for cooling belts ($kg \cdot m^{-2} \cdot hr^{-1}$), only the upper belt area is considered. Strongly adhering pastilles will lead to dust formation at the discharge point.

3.4. THREE TYPICAL APPLICATIONS

A case of pastillation of specialty waxes has been reported (Sandvik, 1990). The specialty waxes are used in cosmetics, pharmaceuticals, and candle production. Pastillation replaces the production of 10-lb slabs because precise control of product weight is difficult in the processing of waxes into slabs. On pastillating, molten wax is pumped into the dropformer. The resulting pastilles are 3–5 mm in diameter.

The production of sulfur/bentonite pastilles for an agricultural application has been described (Schuster and Hodel, 1990). Having sulfur in fertilizer is important to growing plants. For sulfur to be available to plants, it must be oxidized to a water-soluble sulfate.

The pastilles produced are hard, uniform, free-flowing, and virtually dust-free. In contact with moisture, the bentonite within the sulfur/bentonite particle swells and this results in the fractionation or disintegration of the particle. Pastilles compare favorably with sulfur prills, the prior product form. Sulfur prills contain dust, presenting a potentially explosive hazard.

In plants, liquid sulfur is mixed with bentonite and the suspension is pastillated by means of a dropformer. The dropformer contains an internal agitator to prevent the bentonite from settling before solidification. The abrasive nature of the bentonite requires special attention to select suitable materials of construction for some of the rotary dropformer components.

The production of 25-kg bitumen blocks has been reported (Sandvik, 1980). Oxidized bitumen is used in small amounts for roofing and grouting on building sites. The 25-kg blocks are wrapped in polyethylene. On site, the blocks, together with their polyethylene wrapping, are thrown directly into the melting vessel. The wrapping material melts and mixes with the bitumen, thereby eliminating the need to remove the packaging material.

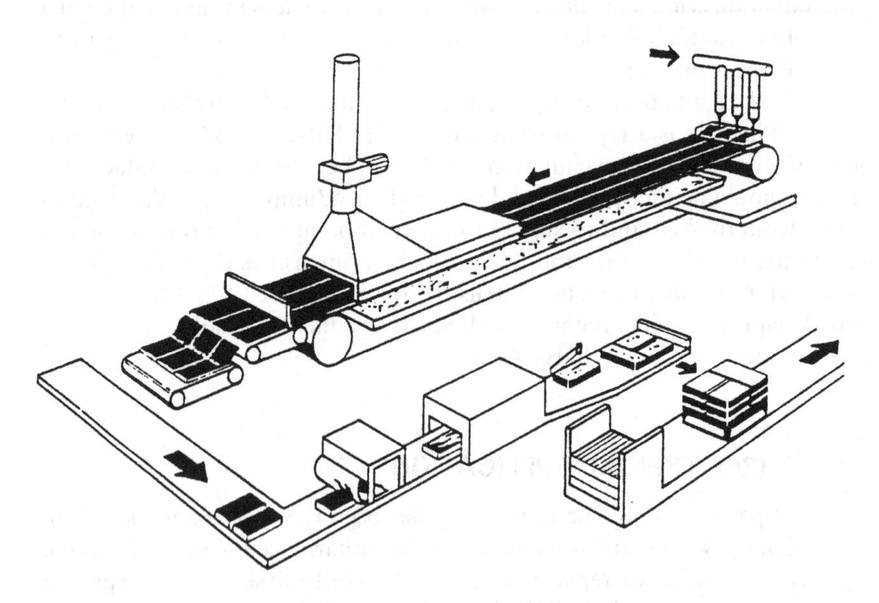

Bitumen System Nynäs

FIG. 3.5 The production of bitumen blocks. (Courtesy of Sandvik Process Systems, Fellbach/Stuttgart, Germany.)

The production proceeds by loading the liquid bitumen onto a belt cooler by means of three weir overflow feeders in parallel (Fig. 3.5). Three separate lanes are obtained, each having a width of 345 mm. The 10-mm-thick layer can be cooled in less than 5 min. The resultant sheets are cut into slabs having a length of 710 mm. Ten slabs are stacked on top of each other to form a bitumen block having a standard $710 \times 345 \times 100$ mm^3 size.

The setup replaces filling paper bags, cardboard drums, etc. with liquid bitumen. The latter procedure has many disadvantages, including that it is labor-intensive and environmentally unsatisfactory for those involved. The packaging material is expensive and logistical problems arise due to the very long cooling times.

3.5. PASTILLATION

Plungers and nozzles were used exclusively before the rotary dropformers were introduced. They came on the market in the 1950s, whereas the rotary dropformers became available in the 1980s. As stated in Sec. 3.2, rotary dropforming equipment permits larger capacities per meter of belt width than plungers and nozzles. The explanation is as follows. The plungers and nozzles that form the drops are stationary while the belt moves. On surpassing a certain velocity, the deposited droplets tend to deform into tears. Of course, this velocity depends on the full set of independent process variables mentioned in Sec. 3.3 and can be established experimentally.

The circumferential velocity of a rotating dropformer can be matched to the linear belt velocity. Thus, there is, in principle, no velocity difference between the rotary dropformer and the belt. The consequence is that, with rotary dropformers, higher belt velocities can be achieved than when using plungers and nozzles. This is a bonus because it is relatively easy to make the belts as long as necessary to provide the required cooling time.

With plungers and nozzles, the drops are deposited under gravity. With rotary dropformers, the drop is deposited under gravity; however, there is also a centrifugal component—albeit this component is small. Whereas the acceleration due to gravity is $9.81 \text{ m} \cdot \text{sec}^{-2}$, the radial acceleration of a rotary dropformer having a diameter of 8 cm and a circumferential velocity of $10 \text{ m} \cdot \text{min}^{-1}$ is $0.07 \text{ m} \cdot \text{sec}^{-2}$.

The production of pastilles implies less efficient belt utilization than the production of flakes. Furthermore, the thickness of flakes is an independent process variable, whereas the height of a pastille is related to the diameter of the pastille.

On pastillating, the most efficient utilization of the belt area is obtained when the pastilles are deposited at the corners of an equilateral triangle. It is normal to allow for a fixed space of 3 mm between the pastilles. This will prevent the pastilles from joining and thus forming "bicycle chains". The impact of the 3-mm space regarding efficient belt utilization will be greater for small pastilles than for large pastilles. On the other hand, large pastilles need more time to solidify because of their larger height.

Example

The deposition of 5-mm pastilles is compared to the deposition of 10-mm pastilles.

The pastilles have a hemispherical shape.
The distance between the pastilles is 3 mm.
The specific mass of the pastilles is 1200 kg $\cdot$ m^{-3}.

5-mm pastilles
Area of the equilateral triangle: $0.5 \cdot 8 \cdot 4\sqrt{3} = 27.7$ mm^2
Area covered with pastilles: $0.5\ (\pi/4)\ 5^2 = 9.8$ mm^2
Area utilization: $(9.8/27.7)\ 100 = 35.4\%$
Time for solidification and cooling: 60 sec
Material on the belt
Weight of one pastille: $0.5(\pi/6)5^3 \cdot 1200 \cdot 10^{-9} = 3.93 \cdot 10^{-5}$ kg

$$\frac{0.5 \cdot 3.93 \cdot 10^{-5}}{27.7 \cdot 10^{-6}} = 0.709\ \text{kg} \cdot \text{m}^{-2}$$

Production intensity: $60 \cdot 0.709 = 42.5$ kg $\cdot$ m^{-2} $\cdot$ hr^{-1}

10-mm pastilles
Area of the equilateral triangle: $0.5 \cdot 13 \cdot 6.5\sqrt{3} = 73.2$ mm^2
Area covered with pastilles: $0.5(\pi/4)10^2 = 39.3$ mm^2
Area utilization: $(39.3/73.2) \cdot 100 = 53.7\%$
Time for solidification and cooling: 240 sec (approximately $2^2 \cdot 60$)
Material on the belt
Weight of one pastille: $0.5(\pi/6)10^3 \cdot 1200 \cdot 10^{-9} = 3.14 \cdot 10^{-4}$ kg

$$\frac{0.5 \cdot 3.14 \cdot 10^{-4}}{73.2 \cdot 10^{-6}} = 2.14\ \text{kg} \cdot \text{m}^{-2}$$

Production intensity: $15 \cdot 2.14 = 32.1$ kg $\cdot$ m^{-2} $\cdot$ hr^{-1}

Conclusion

Although the belt is utilized better when covered with larger pastilles, this effect is counteracted by the longer solidification and cooling times for higher pastilles.

Sandvik's Rotoform consists of a fully heated, stationary inner part, called the stator, surrounded by a rotating perforated shell, called the rotor (Fig. 3.6). The filtered feed enters via a cylindrical feed pipe and is subse-

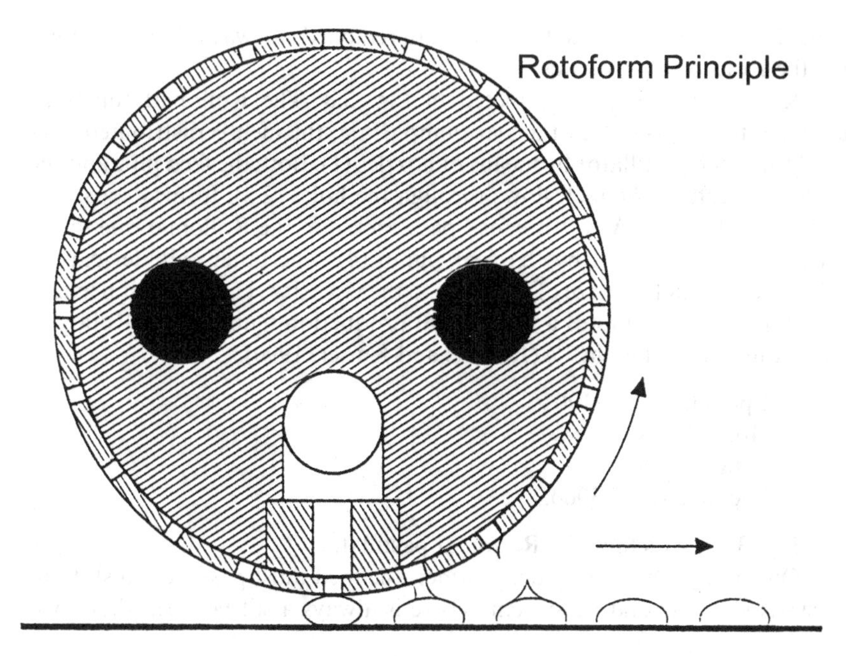

FIG. 3.6 Rotary dropformer. (Courtesy of Sandvik Process Systems, Fellbach/Stuttgart, Germany.)

quently distributed to a metering bar. A longitudinal slot in the stator, in conjunction with the metering bar, ensures that a small portion of molten material is deposited on the belt each time the slot is lined up with holes in the shell. The metering bar evens out the pressure of the molten material (normally 2–5 bar, maximum 20 bar). This bar is pressed against the rotor by means of springs when melts having a low viscosity are processed. The metering bar for melts having a high viscosity is fixed in the stator by means of cams. Now, there is a slit between the rotor and the bar. The heating medium circulates through the channels shown in black. Auxiliary heating by means of electricity is possible. It is also possible to install the Rotoform in a heated hood. This all points to the need to thermostat carefully. The overspill on the outside of the rotor and in the holes of the rotor is sucked back into the stator by a special construction. This construction consists of a cavity between the stator and the rotor and an external refeed bar. The Rotoform has a diameter of 80 mm and the maximum length is 1500 mm. The diameter and pattern of the holes are important independent process variables. It is possible to exchange a rotor by a different rotor, and it is necessary to have room available for these switches; the rotors are removed sideways. Fig. 3.7 shows how the Rotoform pastillates.

Kaiser's Rollomat is depicted in Fig. 3.8. The Rollomat functions according to the principle of a gear pump. First, the melt is distributed over the width of the pastillator by means of nozzles (1). Both the feed pipe and the nozzle are heated. The melt flows through the holes (2,3), forms drops, and solidifies on the belt. Any "threads" pass back into the Rollomat due to the underpressure generated by the gear pump (4). The outer cylinder can be exchanged to match the size of the holes to the viscosity. The maximum length is 1500 mm. Kaiser quotes a few typical production figures for a Rollomat having this length ($kg \cdot hr^{-1}$) (Robens and Kaiser, 1990):

- Epoxide resin: 3100
- Hotmelt: 1800
- Paraffin: 1200
- Phenolic resin: 3000.

Fig. 3.9 shows how the Rollomat pastillates.

On employing a rotary dropformer, the melt must pass from a stationary system into a rotating system. There is always a slit between these two

FIG. 3.7 Rotoform in operation. (Courtesy of Sandvik Process Systems, Fellbach/Stuttgart, Germany.)

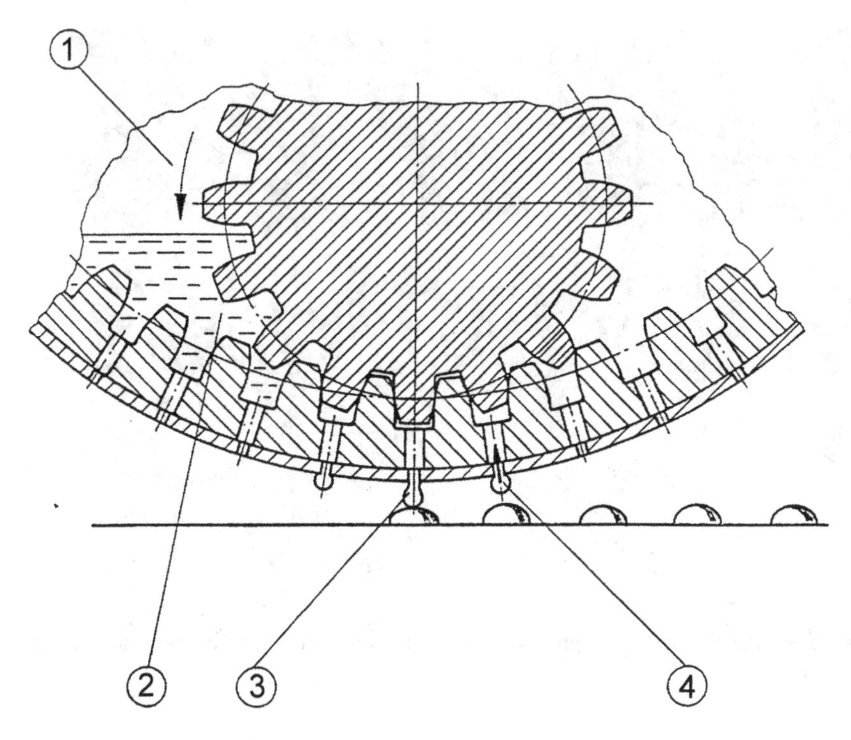

FIG. 3.8 Rotary dropformer. (Courtesy of Kaiser, Krefeld, Germany.)

systems and one must install the usual provisions to prevent the melt from leaking out (e.g., a single mechanical seal).

Goudsche Machinefabriek introduced the Disc Pastillator recently. Fig. 3.10 shows how the pastillator works. Fig. 3.11 shows a Disc Pastillator having two discs. Like a plate dryer, the Disc Pastillator contains one or more horizontal jacketed discs. The discs are attached to a shaft that rotates. There is a feeding device for each disc and these feeding devices make a reciprocating movement. For a certain angle, they move along with the discs at the same angular speed. The droplets are deposited during this stroke. The deposition occurs under gravity and, because the movements of the discs and the feeding devices are synchronized, the pastilles are shaped well. At the end of this stroke, the feeding devices move back and a new cycle starts. A sector of a disc is covered with pastilles at each stroke. The pastilles are removed by means of a scraper when they have made almost one complete turn. This is shown in Fig. 3.12. A cooling medium circulates through the jackets of the discs. Contamination of the cooling medium with the product and vice versa is impossible because there is a physical barrier between the droplets/pastilles

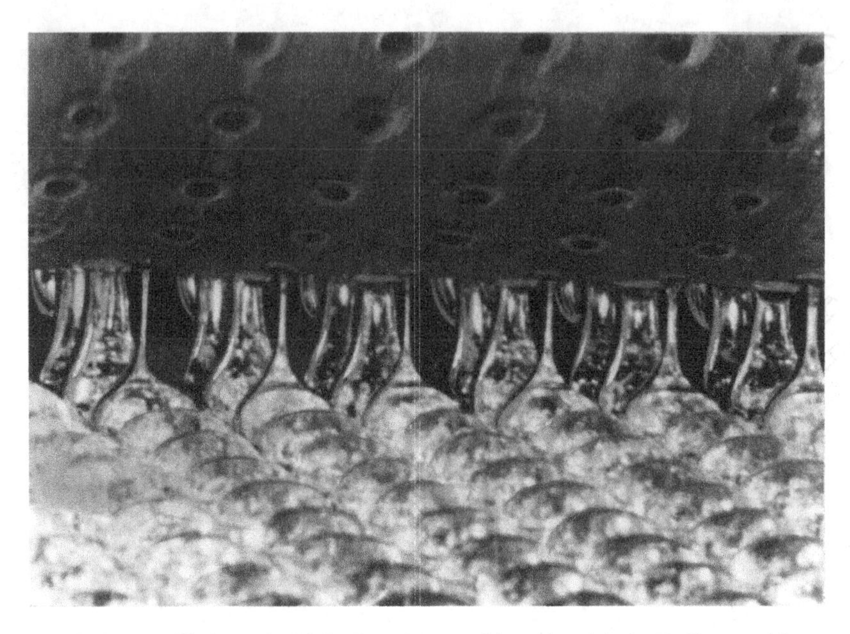

FIG. 3.9 Pastillation of a high-viscosity epoxide resin with the Rollomat. (Courtesy of Kaiser, Krefeld, Germany.)

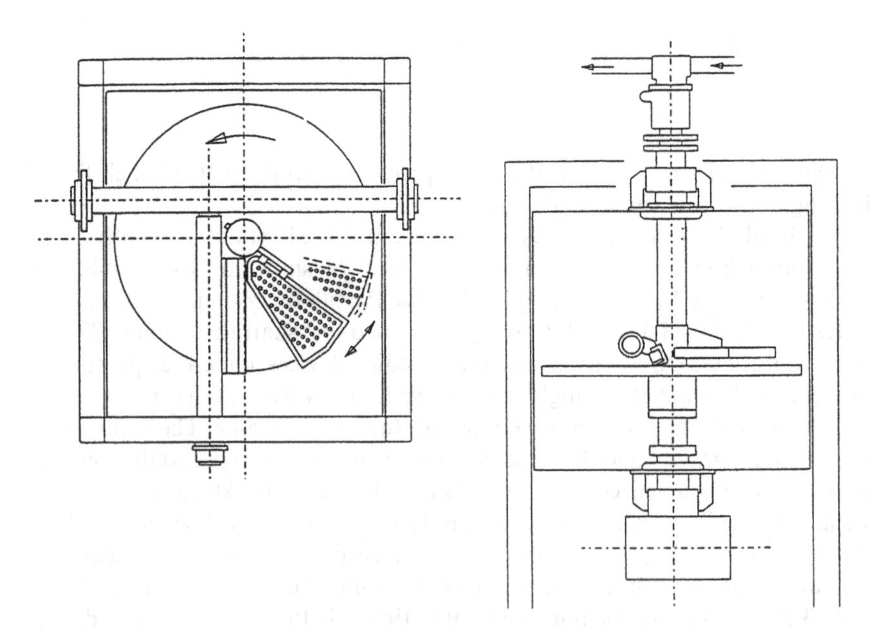

FIG. 3.10 Schematic drawing of the Disc Pastillator. (Courtesy of Goudsche Machinefabriek, Waddinxveen, The Netherlands.)

FIG. 3.11 A Disc Pastillator having two discs. (Courtesy of Goudsche Machinefabriek, Waddinxveen, The Netherlands.)

and the cooling medium. Coolants other than water can be used because the coolant is circulated in a closed system (such as in a drum flaker). Condensation of water from the air on the discs can be prevented by admitting dry air or nitrogen to the casing of the Disc Pastillator. The maintenance of a certain gas atmosphere is technically feasible because the volume of even a large pastillator is relatively small.

The cooling medium is fed and discharged through the hollow shaft.

The smallest Disc Pastillator contains one disc having an area of 0.74 m^2, whereas the largest pastillator contains six discs, each having an area of 3 m^2.

The rotational speed of the shaft is continuously variable. One revolution per minute will give a residence time of 1 min, 2 rpm results in a residence time of 0.5 min, and so on.

FIG. 3.12 The Disc Pastillator in operation. (Courtesy of Goudsche Machinefabriek, Waddinxveen, The Netherlands.)

The main field of application of this pastillator will be in fine chemicals and pharmaceuticals. The good control of process conditions is a bonus for this dropformer. Large and very large belt coolers can do with one feeding device, whereas a large Disc Pastillator needs a number of feeding devices.

The feasibility of using this dropformer can be checked by means of hand tests. Droplets can be deposited on a cold stainless steel plate by means of a pipette. The next step is testing the smallest Disc Pastillator. Scaling up can occur by means of the registered production intensity ($kg \cdot m^{-2} \cdot hr^{-1}$).

Steel Belt Systems introduced a rotary dropformer recently. This rotary dropformer is called the Rolldrop and it consists of three main parts:

- A feeding device
- A cylindrical roll carrying teeth on its outer surface
- A feed distributor.

The feeding device is a jacketed cylinder, which is as long as the toothed roll and sits on top of the latter roll. The melt is fed to the feeding device and leaves this cylinder through a slot at the bottom. It flows directly on the toothed roll. The latter roll is thermostatted by an internal circulation. It rotates and the melt drips from the teeth onto the moving belt. A feed

FIG. 3.13 Rotary dropformer. (Courtesy of Steel Belt Systems, Milan, Italy.)

distributor passes the melt from the space between the teeth to the teeth. It is constructed as a comb and it is stationary. The feed distributor is also jacketed. The standard lengths of the cylindrical roll are 600, 800, 1000, 1200, and 1500 mm. Wax having a viscosity in the range of 200–300 mPa · sec was processed successfully. A viscosity of 20 Pa · sec has also been tested. Fig. 3.13 shows the pastillator. The Rolldrop is a relatively simple pastillator that is easily accessible for maintenance and cleaning.

3.6. MISCELLANEOUS

Belts are made of carbon or stainless steel. Table 3.1 contains physical data of belt materials. Standard widths for stainless steel AISI 316 belts are (in millimeters): 600, 800, 1000, 1200, 1400, 1500, 1560, 3000, and 4500. A belt having a width of 3000 mm is obtained by combining two belts, each having a width of 1500 mm, by means of a longitudinal weld. Likewise, a width of 4500 mm is obtained by means of two longitudinal welds. Cooling belts can be up to 20 m long or even longer. The length of a cooling module is, for example, 2.5 m. The standard thickness of stainless steel AISI 316 belts is 1.0 mm. As a rule-of-thumb, the drum diameter exceeds the belt thickness by a factor of 1000.

Productions up to 10 t · hr^{-1} per belt are possible. Production rates up to 100 kg · m^{-2} · hr^{-1} are obtained for flakes, whereas this figure is 50 for

TABLE 3.1 Physical Properties of Belt Materials

Material type	Carbon steel	Stainless steel
Belt grade	1300C	1000SA
AISI number	1065	316
"Werkstoff" number	1.1235	1.4436
Specific mass [$kg \cdot m^{-3}$]	7850	8000
Thermal conductivity [$W \cdot m^{-1} \cdot K^{-1}$]		
20°C	42	15
100°C	38	16
200°C	35	18
Specific heat [$kJ \cdot kg^{-1} \cdot K^{-1}$]		
50–100°C	0.50	0.50
250–300°C	0.54	0.54

Source: Helber (1993).

pastilles. The production rate for a given product depends strongly on the melting point. High capacities are possible for products having a high melting point and vice versa.

The production of micropastilles, having a diameter as small as 1 mm, is a recent Sandvik development. With this new technique, the distance between the belt and the rotary dropformer is so small that the product, when it protrudes from the bores in the form of a wart, is dragged along by the belt.

Sometimes, on producing pastilles, "twins" or even "bicycle chains" are observed. The pastilles join up. This may be traced back to a hole pattern that is too critical for the application, imperfect temperature control, uneven pressure distribution, or an imbalance of the rotary dropformer.

A double-belt cooler can be used if it is desired to produce flakes. In a double-belt cooler, the material is processed between two horizontal steel belts and this guarantees good belt contact (Fig. 3.14).

Some melts suffer from cracking or decomposition. It is recommended to check this by means of storage tests.

Air circulation over sheet materials or pastilles gives additional cooling. Thus, there can be a solidification front moving top–down and one moving bottom–up. However, the coefficients for the heat transfer solid material/air are low. Furthermore, the solidified material may have a weak plane where the two moving solidification fronts have met.

Some operational data:

- Belt velocities are in the range of 0.1–60 $m \cdot min^{-1}$.
- Residence times vary from seconds to minutes.
- Melt temperatures are in the range of 30–300°C.

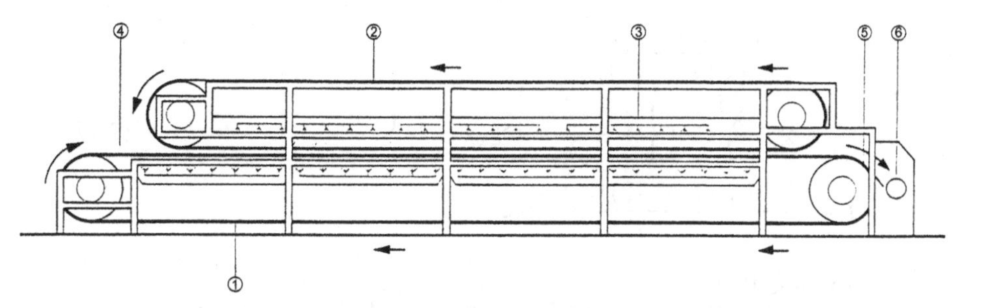

FIG. 3.14 Double-belt cooler. The product is fed onto the upper surface of the lower belt (1), which carries it into the central cooling zone, where the pressure of the upper belt (2) ensures constant contact with both cooling surfaces. The upper belt is sprayed with cooling water (3) from above, and the lower belt from below. Loading (4) and discharge provisions (5) are arranged to suit individual products, and can incorporate breaker equipment (6) at the discharge station if required. (Courtesy of Sandvik Process Systems, Fellbach/Stuttgart, Germany.)

3.7. THE DESIGN METHOD FOR BELT COOLERS

First, solidification and cooling should be distinguished. Solidification occurs on a wall with a variable temperature, whereas cooling takes place on a wall with a fixed temperature (i.e., the cooling water temperature). It is recommended to consider the solidification as proceeding according to Case 3 of Sec. 2.5. There, the situation—that the melt has a temperature higher than its melting point and is poured on a metal surface while all heat transfer is to the metal surface—is addressed. The horizontal metal surface is cooled by spraying cooling water against the underside. The theory for the cooling of the solid layer is discussed in Sec. 2.7. Both the temperature of the surface and the average temperature are dealt with.

Material balance

$$\phi_p = \delta B v_b \rho_s \ \text{kg} \cdot \text{sec}^{-1}$$

Heat balance

$$Q = \phi_p\{c_l(T_f - T_o) + i + c_s(T_o - \overline{T})\}\ \text{W}$$

Cooling water flow

$$\phi_w = \frac{Q}{4200\,\Delta T_c}\ \text{kg} \cdot \text{sec}^{-1}$$

Solidification time

$$\chi = \frac{\alpha_k \delta}{\lambda_s}$$

$$\mu' = \frac{c_s(T_o - T_c)}{2\{i + c_l(T_f - T_o)\}}$$

$$\tau' = \chi + \frac{\chi^2}{2} + \mu'\left\{\chi + \frac{\chi^2}{2} - {}^e\log(\chi + 1)\right\} \tag{2.17}$$

$$t = \frac{\tau'\{i + c_l(T_f - T_o)\}\rho_s\lambda_s}{\alpha_k^2(T_o - T_c)} \text{ sec}$$

Cooling time

The temperature T of the distant (insulated) face is given by:

$$(T - T_o) = \sum_{N=0}^{N=\infty}(-1)^N(T_w - T_o)2 \cdot \text{erfc}\frac{(2N+1)h}{2\sqrt{at}} \text{ K} \tag{2.23}$$

In many instances, the second and higher terms are negligible compared with the first term. This can be established by inspection.

The average temperature $\overline{T}$ can be assessed as follows:

$$\frac{T_w - \overline{T}}{T_w - T_o} = \frac{8}{\pi^2}\left(e^{-bFo} + \frac{1}{9}e^{-9bFo} + \frac{1}{25}e^{-25bFo} + \cdots\right) \tag{2.24}$$

In many instances, only the first term of the series is significant.

Equation (2.25) is a simplification of Eq. (2.24). It is applicable when $Fo > 0.1$:

$$t = \frac{1}{a}\left(\frac{2h}{\pi}\right)^2 {}^e\log\frac{8(T_w - T_o)}{\pi^2(T_w - \overline{T})} \tag{2.25}$$

3.8. VERIFICATION OF THE DESIGN METHOD

3.8.1. Small-Scale Trials with a Product (from Reith, 1976)

Physical Properties

$T_o = 61°C$
$i = 138{,}000 \text{ J} \cdot \text{kg}^{-1} \cdot \text{K}^{-1}$
$\rho_s = 1360 \text{ kg} \cdot \text{m}^{-3}$ (25°C)
$c_l = 1880 \text{ J} \cdot \text{kg}^{-1} \cdot \text{K}^{-1}$ (80°C)
$c_s = 1610 \text{ J} \cdot \text{kg}^{-1} \cdot \text{K}^{-1}$ (25°C)
$\lambda_s = 0.26 \text{ W} \cdot \text{m}^{-1} \cdot \text{K}^{-1}$ (25°C)

Equipment

A horizontal, circular, stainless steel plate with a rim having a diameter of 275 mm and a thickness of 4 mm was used. The cooling medium was sprayed.

$\lambda_w = 14.5\ W \cdot m^{-1} \cdot K^{-1}$

α_k

$$\frac{1}{\alpha_k} = \frac{1}{\alpha_c} + \frac{\delta_w}{\lambda_w}\ W^{-1} \cdot m^2 \cdot K$$

$\alpha_c = 1400\ W \cdot m^{-2} \cdot K^{-1}$ (assumption)

For the calculations, $\alpha_k = 1000\ W \cdot m^{-2} \cdot K^{-1}$.

General Description

A series of trials comprised 16 experiments. Table 3.2 contains the experimental and the calculated data. The product was of technical grade. The melt of temperature T_f (°C) was poured on the plate and the crystallization time was determined visually using a stopwatch. For each of the series of 16 experiments, the crystallization time was determined three times. The three measurements were arithmetically averaged. The difference between these three measurements was a maximum of 3%.

Ranges of Independent Variables

T_c: −10°C to 32°C
T_f: 70°C and 90°C
δ_{max}: 2.9–4.3 mm

Ranges of Dependent Variables

t_{act}: 87–335 sec

Description of Crystallization

This covers the crystallization on the plate. The approach recommended in Sec. 3.7 was used to calculate crystallization times and these times were compared to the actual crystallization times. The predictions were, on average, 2% greater than the actual. In other words, $t_{pred}/t_{act} = 1.02$. The ratios were in the range of 0.88–1.20 (Table 3.2).

Additional Remarks

The cooling process was not checked during this series of trials. In this series of trials, the heat transfer rate was determined by the resistance offered by the crystallized layer. In other words, the impact of a greater α_k value is

TABLE 3.2 Data for Small-Scale Belt Cooler Trials

Experiment Number	1	2	3	4	5	6	7	8	9	10	11	12	13	14	15	16
Independent variables																
δ_{max} [mm]	3.0	4.3	3.1	3.1	2.9	3.0	3.6	4.1	3.5	3.4	3.1	3.6	3.5	3.45	3.0	2.9
T_f [°C]	90	90	90	70	70	70	90	90	90	70	90	90	90	90	90	70
T_c [°C]	32	32	32	32	32	32	21	20	15	15	10	7	−1	−1	−10	−10
Dependent variable																
t_{act} [sec]	195	335	195	180	147	165	180	300	175	150	152	165	138	144	107	87
Results of calculations																
t_{pred} [sec]	205	403	218	180	159	169	212	272	183	145	134	169	143	139	96	77
t_{act}/t_{pred}	1.05	1.20	1.12	1.00	1.08	1.02	1.18	0.91	1.04	0.97	0.88	1.02	1.04	0.97	0.90	0.88

only marginal. Specifically, the crystallization times predicted when $\alpha_k = 10{,}000 \text{ W} \cdot \text{m}^{-2} \cdot \text{K}^{-1}$ are only 10–15% shorter than the times calculated when $\alpha_k = 1000 \text{ W} \cdot \text{m}^{-2} \cdot \text{K}^{-1}$. It is improbable that α_k was smaller than $1000 \text{ W} \cdot \text{m}^{-2} \cdot \text{K}^{-1}$.

3.8.2. Large-Scale Trials with a Product (from Reith, 1976)

Physical Properties

See Sec. 3.8.1

Equipment

A stainless steel belt cooler having a length of 15 m, a width of 1.09 m, and a thickness of 2 mm was used. $\lambda_w = 14.5 \text{ W} \cdot \text{m}^{-1} \cdot \text{K}^{-1}$. The cooling water was sprayed and the belt speed was adjustable.

α_k

$$\frac{1}{\alpha_k} = \frac{1}{\alpha_c} + \frac{\delta_w}{\lambda_w} \quad \text{W}^{-1} \cdot \text{m}^2 \cdot \text{K}$$

$\alpha_c = 1200 \text{ W} \cdot \text{m}^{-2} \cdot \text{K}^{-1}$ (assumption).
For the calculations, $\alpha_k = 1000 \text{ W} \cdot \text{m}^2 \cdot \text{K}^{-1}$.

General Description

The trials consisted of six experiments. A technical quality of the product was used. A defined melt flow was poured on the belt moving at a certain speed. The length needed for complete crystallization was determined visually. In two experiments, the average flake temperature was measured. The average flake thickness was determined in each experiment.

Description of Crystallization

Table 3.3 contains the data of the six experiments. First, the material balances of the experiments were checked by means of the material balance equation of Sec. 3.7. Experiment 2 had to be excluded. The approach recommended in Sec. 3.7 was used to calculate crystallization times and these times were compared to the actual crystallization times. It appears that, on the average, the actual crystallization times are 1.3 times larger than the calculated crystallization times. Probably, the reasons are as follows:

- The adhesion to the moving stainless steel belt is less perfect than the adhesion at the tests described in Sec. 3.8.1.
- The assessments of the lengths needed for crystallization are inaccurate.

TABLE 3.3 Data for Large-Scale Belt Cooler Trials

Experiment Number	1	2	3	4	5	6
Independent variables						
ϕ_P [kg · hr^{-1}]	709	709	1170	878	954	792
v_b [m · min^{-1}]	2.5	2.25	3.7	3.1	2.9	2.8
T_f [°C]	86	86	79	90	88	90
T_c [°C]	28.8	24.5	19.0	18.5	16.0	18.0
Dependent variables						
h [mm]	3.05	2.21	3.36	3.22	3.64	3.19
L_{c1} [m]	10.5	10.5	11.5	13.0	13.5	8.0
L_{c2} [m]	4.5	4.5	3.5	2.0	1.5	7.0
$\overline{T}$ [°C]	30	—	25.3	—	—	—
Results of calculations						
Crystallization						
t_{pred} [sec]	186		166	167	197	163
t_{act} [sec]	251		186	248	280	174
t_{act}/t_{pred}	1.35		1.12	1.49	1.42	1.07
Cooling						
t_{pred} [sec]	98		65			
t_{act} [sec]	108		57			
t_{act}/t_{pred}	1.10		0.88			

- For a given experiment, the layer thickness varies (standard deviations of approximately 0.5 mm were found).

For this product, it is recommended to multiply calculated industrial crystallization times and lengths by 1.3.

Description of Cooling

The cooling times were checked in Experiments 1 and 3. In both cases, the accuracy of the predictions was ±10%.

Additional Remark

The observations regarding α_k in Sec. 3.8.1 are also applicable here.

3.9. EXAMPLE OF FLAKE PRODUCTION

Designing a belt for the solidification and cooling of 3 t · hr^{-1} naphthalene is requested. Flakes having a thickness of 3 mm are desired. The full belt width is 1560 mm, whereas the effective belt width is 1500 mm. The belt is made of stainless steel AISI 316 and has a thickness of 1 mm. The average cooling water temperature is 15°C. The feed is at 90°C. The flakes should have a maximum surface temperature of 30°C. The average product temperature should be 20°C.

Physical Data of Naphthalene

T_o = 80.4°C
i = 148,100 J · kg^{-1}
ρ_s = 1150 kg · m^{-3} (60°C)
c_l = 1710 J · kg^{-1} · K^{-1}
c_s = 1440 J · kg^{-1} · K^{-1} (51°C)
λ_s = 0.37 W · m^{-1} · K^{-1} (38°C) (Groot Wassink, 1976)

Belt Velocity

$$\frac{3000}{3600} = 1.5 \cdot 0.003 \cdot 1150 \cdot v_b \rightarrow v_b = 0.161 \text{ m} \cdot \text{sec}^{-1}$$

Heat Transferred

Melt cooling	3000 · 1.710(90–80.4)	= 49,248
Solidification	3000 · 148.100	= 444,300
Solid cooling	3000 · 1.440(80.4–20)	= 260,928 +
		754,476 kJ · hr^{-1}
		= 209.6 kW

Cooling Water Flow

The cooling water temperature rises from 13°C to 17°C.
$\phi_w \cdot 4.2\ (17 - 13) = 754{,}476 \rightarrow \phi_w = 44{,}909$ kg · hr^{-1} (i.e., 45 m^3 · hr^{-1})

Solidification Time

Incidentally, the layer thickness can be 3.5 mm.

$$t = \left[\frac{\{i + c_l(T_f - T_o)\}\rho_s}{2(T_o - T_w)\lambda_s} + \frac{\rho_s c_s}{4\lambda_s}\right]\delta^2 \tag{2.5}$$

$$t = \left[\frac{\{148,100 + 1710(90 - 80.4)\}1150}{2(80.4 - 15)0.37} + \frac{1150 \cdot 1440}{4 \cdot 0.37}\right]3.5^2 \cdot 10^{-6}$$

$$= 61.6 \text{ sec}$$

The assumption is that the cooling belt is at 15°C throughout.

$$\tau' = \chi + \frac{\chi^2}{2} + \mu'\left\{\chi + \frac{\chi^2}{2} - {}^{e}\log(\chi + 1)\right\} \tag{2.17}$$

$\alpha_c = 1000 \text{ W} \cdot \text{m}^{-2} \cdot \text{K}^{-1}$ (assumption)
$\lambda_w = 15 \text{ W} \cdot \text{m}^{-1} \cdot \text{K}^{-1}$ (stainless steel)

$$\frac{1}{\alpha_k} = \frac{1}{\alpha_c} + \frac{\delta_w}{\lambda_w} = \frac{1}{1000} + \frac{0.001}{15} = 0.001066$$

$$\alpha_k = 935 \text{ W} \cdot \text{m}^{-2} \cdot \text{K}^{-1}$$

$$\chi = \frac{\alpha_k \delta}{\lambda_s} = \frac{938 \cdot 0.0035}{0.37} = 8.87$$

$$\mu' = \frac{c_s(T_o - T_c)}{2\{i + c_l(T_f - T_o)\}} = \frac{1440(80.4 - 15)}{2\{148,100 + 1710(90 - 80.4)\}} = 0.286$$

$$\tau' = 8.87 + \frac{8.87^2}{2} + 0.286\left(8.87 + \frac{8.87^2}{2} - {}^{e}\log 9.87\right) = 61.34$$

$$t = \frac{\tau'\{i + c_l(T_f - T_o)\}\rho_s\lambda_s}{\alpha_k^2(T_o - T_c)}$$

$$= \frac{61.34\{148,100 + 1710(90 - 80.4)\}1150 \cdot 0.37}{935^2(80.4 - 15)} = 75.1 \text{ sec}$$

Now, the resistances to heat transfer of the cooling water film and the belt are not neglected.

For drum flakers, the spraying of cooling water against the cylinder wall causes heat transfer coefficients in the range of 1000–3000 $\text{kcal} \cdot \text{m}^{-2} \cdot \text{hr}^{-1} \cdot \text{K}^{-1}$ (Preger, 1969). The solidification time becomes 62.9 sec when α_k = 10,000 $\text{W} \cdot \text{m}^{-2} \cdot \text{K}^{-1}$.

Solidification time: 1.1 · 75.1 = 82.6 sec (10% margin)
Belt length: 82.6 · 0.161 = 13.30 m (for solidification)

Cooling Time

$$(T - T_o) = \sum_{N=0}^{N=\infty} (-1)^N (T_w - T_o) 2 \cdot \text{erfc}\frac{(2N+1)h}{2\sqrt{at}} \text{ K} \tag{2.23}$$

$$a = \frac{\lambda_s}{c_s \rho_s} = \frac{0.37}{1440 \cdot 1150} = 2.23 \cdot 10^{-7} \text{ m}^2 \cdot \text{sec}^{-1}$$

$$(30 - 80.4) = (15 - 80.4)2 \cdot \text{erfc}\frac{3.5 \cdot 10^{-3}}{2\sqrt{2.23 \cdot 10^{-7} \cdot t}} - (15 - 80.4)2 \cdot \text{erfc}\frac{3 \cdot 3.5 \cdot 10^{-3}}{2\sqrt{2.23 \cdot 10^{-7} \cdot t}} + \cdots$$

$t = 38$ sec (use Table 2.1)

Inspection shows that the third term can be neglected.

The average flake temperature is found by applying Eq. (2.24). It follows that $\overline{T} = 25°\text{C}$.

$$Fo = \frac{2.23 \cdot 10^{-7} \cdot 38}{(3.5 \cdot 10^{-3})^2} = 0.692$$

$$\frac{15-25}{15-80.4} \approx \frac{8}{\pi^2}\left(e^{-2.467 \cdot 0.692} + \frac{1}{9}e^{-9 \cdot 2.467 \cdot 0.692} + \cdots\right)$$

$$0.1529 \approx \frac{8}{\pi^2}(0.1814 + 2.361 \cdot 10^{-8} + \cdots)$$

Note that this is the average temperature of a flake having a thickness of 3.5 mm.

Belt length (for cooling): $1.1 \cdot 38 \cdot 0.161 = 6.73$ m

Full Belt Length

$13.30 + 6.73 = 20.03$ m

Eight modules of 2.5 m each are required (full length, 20 m).

Residence Time

$$\frac{20}{0.161} = 124 \text{ sec (2 min 4 sec)}$$

Production Intensity

$$\frac{3000}{1.56 \cdot 20} = 96 \text{ kg} \cdot \text{m}^{-2} \cdot \text{hr}^{-1}$$

Remarks

- It is generally recommended to design a minimum cooling length of 5 m for process control reasons.
- The application of the design method requires the availability of the relevant physical data.

- Laboratory experiments and pilot plant tests can confirm the results of the calculations. The possibility of a check of the processing of the material (spreading, adhesion, removal by the scraper, dust, and sublimation) is an additional advantage.
- The results of laboratory experiments and pilot plant tests should be used for design purposes if the physical data are not fully known. The production ($kg \cdot m^{-2} \cdot hr^{-1}$) for a given layer thickness can be used for scaling-up purposes.

3.10. EXAMPLE OF PASTILLE PRODUCTION (1)

The question is raised on how many kilograms of naphthalene pastilles per hour can be made on the cooling belt of the previous example. The diameter of the pastilles is 7 mm, the height is 3 mm, and the pastilles are segments of a sphere. It is safe to assume a solidification time of $0.9 \cdot 61.6 = 55.4$ sec. A cooling time of $0.9 \cdot 38 = 34.2$ sec is also a good assumption. The pastilles are deposited in a triangular pattern. The triangles are equilateral and the length of the sides of the triangle is 9.07 mm. This corresponds with 32 rows of holes in the rotary dropformer, whereas it is assumed that the linear belt velocity equals the circumferential dropformer velocity. Each row of holes contains of 164 holes.

Production

Belt velocity

$$\frac{20}{1.1(55.4 + 34.2)} = 0.203 \text{ m} \cdot \text{sec}^{-1}$$

1.1 is a capacity factor.
Belt coverage (Fig. 3.15).

$$\frac{3 \cdot \frac{1}{6} \cdot \frac{\pi}{4} \cdot 7^2}{0.5 \cdot 9.07 \cdot 0.5 \cdot 9.07 \cdot \sqrt{3}} \cdot 100 = 54.0\%$$

Volume of a spherical segment: $\dfrac{\pi h_1 (3r_1^2 + h_1^2)}{6} = \dfrac{\pi \cdot 3(3 \cdot 3.5^2 + 3^2)}{6}$

$= 71.9 \text{ mm}^3$

Volume of a cylinder having the same height and base:

$$\frac{\pi}{4} \cdot 7^2 \cdot 3 = 115.5 \text{ mm}^3$$

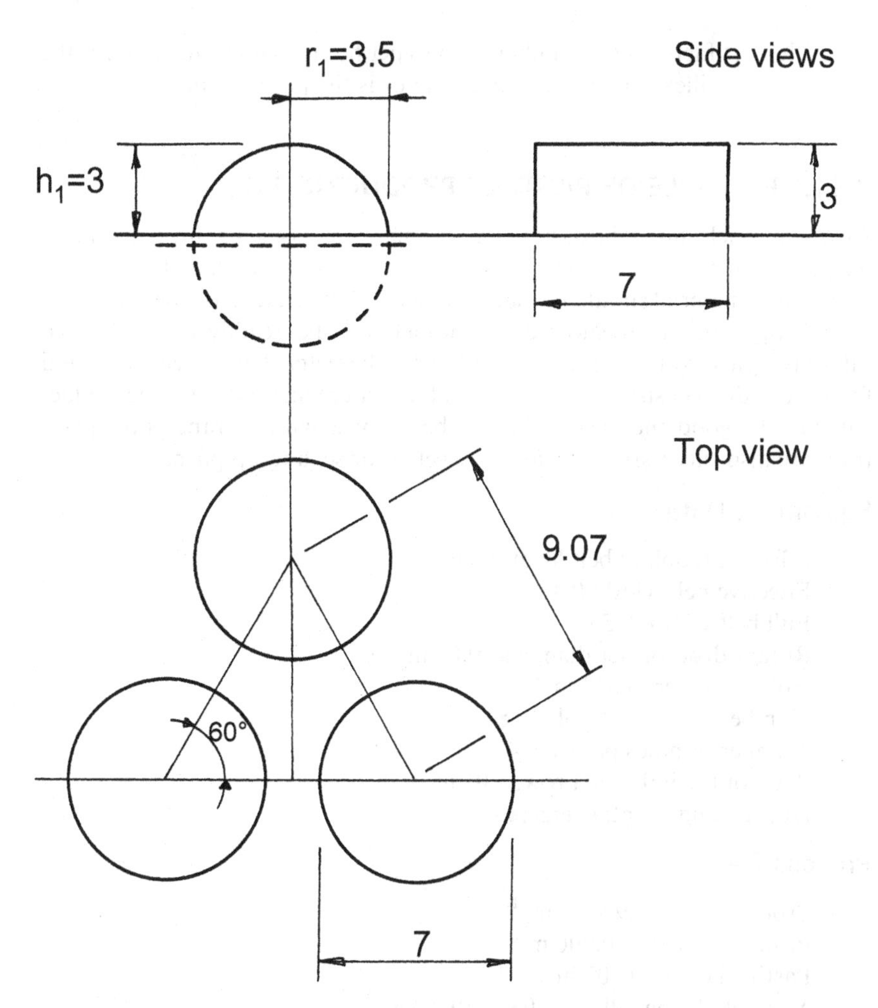

Fig. 3.15 Pastille pattern of the first example on pastillation.

Pastille production

$$\frac{0.203}{0.161} \cdot 0.54 \cdot \frac{71.9}{115.5} \cdot 3000 = 1272 \text{ kg} \cdot \text{hr}^{-1}$$

In practice, the production will be in the range of 1250–1300 kg · hr^{-1} because the pastilles are not very uniform.

Check of the pastille production:

$$\frac{3600 \cdot 0.203}{0.5\sqrt{3} \cdot 9.07 \cdot 10^{-3}} \cdot 164 \cdot 71.9 \cdot 10^{-9} \cdot 1150 = 1262 \text{ kg} \cdot \text{hr}^{-1}$$

The first factor is the number of rows per hour. The second factor is the number of pastilles per row. The third factor is the pastille's mass.

3.11. EXAMPLE OF PASTILLE PRODUCTION (2)

An organic melt should be pastillated. The mass flow is 400 kg $\cdot$ hr^{-1}. The melting point is approximately 60 °C and the viscosity is approximately 1 Pa $\cdot$ sec. The molten material should be blanketed with nitrogen to avoid degradation. Laboratory experiments showed that the melt solidifies readily into round pastilles having a height of approximately half the diameter. This statement is valid for various droplet sizes and various cooling water temperatures. The contact with metal is good and the pastilles can be removed with a scraper. Pilot plant trials are also successful. One test is selected for scaling-up purposes.

Equipment Data

Effective (cooled) belt length: 8 m
Effective belt width: 0.48 m
Full belt width: 0.5 m
Rotary dropformer diameter: 0.08 m
Hole diameter: 0.002 m
Number of rows of holes: 28
Number of holes per row: 47
Pitch of the holes in a row: 0.01 m
Hole arrangement: staggered.

Process Data

Production: 138.9 kg $\cdot$ hr^{-1}
Pastille diameter: 0.006 m
Pastille height: 0.003 m
Mass of 100 pastilles: $4.8054 \cdot 10^{-3}$ kg
Belt speed: 0.133 m $\cdot$ sec^{-1} (8.0 m $\cdot$ min^{-1})
Feed temperature: 66 °C
Product temperature: 32 °C
Cooling water temperature: 20 °C.

Calculation of the Circumferential Velocity of the Dropformer

Mass of 100 pastilles: $4.8054 \cdot 10^{-3}$ kg
Mass of the 47 pastilles produced by one row: $2.2585 \cdot 10^{-3}$ kg
Number of rows produced per hour: $138.9/(2.2585 \cdot 10^{-3}) = 61{,}501$
Row-to-row distance on the belt: $3600 \cdot 0.133/61{,}501$
$= 7.79 \cdot 10^{-3}$ m

Row-to-row distance on the rotary dropformer: $\pi \cdot 0.08/28 = 8.98 \cdot 10^{-3}$ m

Circumferential velocity of the dropformer: $\dfrac{0.133 \cdot 8.98 \cdot 10^{-3}}{7.79 \cdot 10^{-3}}$

$= 0.153 \text{ m} \cdot \text{sec}^{-1}$ $(9.2 \text{ m} \cdot \text{min}^{-1})$

The rotary dropformer rotates supersynchronous.

Belt Utilization/Pastille Pattern

$$\frac{\text{Pastille area produced per hour}}{\text{Belt area traveled per hour}} \cdot 100$$

$$= \frac{\dfrac{138.9}{4.8054 \cdot 10^{-5}} \cdot \dfrac{\pi}{4} \cdot 0.006^2}{8 \cdot 60 \cdot 0.48} \cdot 100 = 35.5\%$$

Fig. 3.16 shows the pastille pattern. There is some room for improvement (i.e., more effective utilization of the area).

Specific Production Capacity

$$\frac{138.9}{0.48 \cdot 8} = 36.2 \text{ kg} \cdot \text{hr}^{-1} \cdot \text{m}^{-2}$$

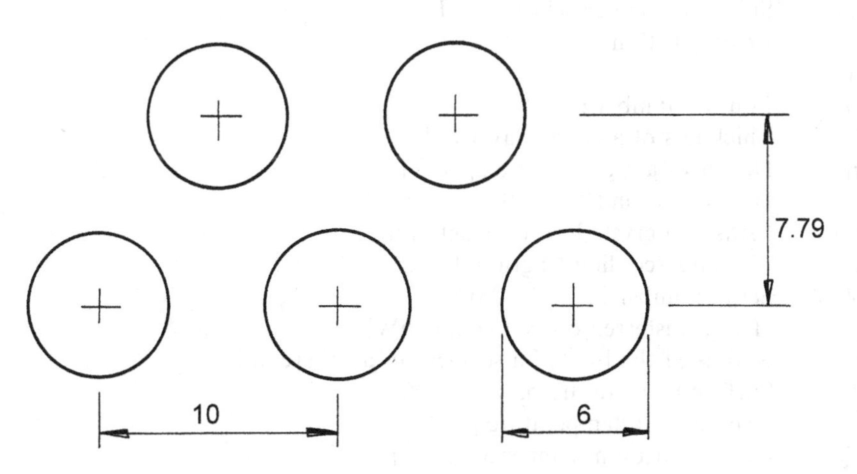

FIG. 3.16 Pastille pattern of the second example on pastillation.

Residence Time

$$\frac{8}{0.133} = 60 \text{ sec}$$

Scaling Up

Commercial area required: $400/36.2 = 11.0 \text{ m}^2$ (e.g., a belt having a width of 1 m and a length of 12.5 m; five modules). The belt velocity is $0.208 \text{ m} \cdot \text{sec}^{-1}$. The rotary dropformer is longer than the one in the pilot plant but has the same hole pattern.

Note 1

If the pitch of the holes in a row is reduced to 8 mm, a 20% capacity gain is possible.

Note 2

It is assumed that the rotary dropformer functions at the higher velocity as it did at the lower (test) velocity.

LIST OF SYMBOLS

a	Thermal diffusivity $\{\lambda_s/(c_s\rho_s)\}$ $[\text{m}^2 \cdot \text{sec}^{-1}]$
B	Cooling belt width [m]
b	$(\pi/2)^2 = 2.467$
c_l	Melt specific heat $[\text{J} \cdot \text{kg}^{-1} \cdot \text{K}^{-1}]$
c_s	Solid specific heat $[\text{J} \cdot \text{kg}^{-1} \cdot \text{K}^{-1}]$
erf	Error function
erfc	1 − erf
Fo	Fourier number (at/h^2)
h	Thickness of a cooled layer [m]
h_1	Height of a segment of a sphere [m]
i	Heat of fusion $[\text{J} \cdot \text{kg}^{-1}]$
L_{c1}	Measured crystallization length [m]
L_{c2}	Measured cooling length [m]
N	Serial number in Eq. (2.23)
Q	Heat transferred on belt cooling [W]
r_1	Radius of the base of a segment of a sphere [m]
T	Surface temperature of a slab [°C]
$\overline{T}$	Average slab temperature [°C]
T_c	Cooling medium temperature [°C]
T_f	Melt temperature [°C]

T_o	Melting point [°C]
T_w	Metal wall temperature (process-side) [°C]
ΔT_c	Temperature increase of the cooling water [K]
t	Time [sec]
t_{act}	Actual crystallization/cooling time [sec]
t_{pred}	Predicted crystallization/cooling time [sec]
v_b	Cooling belt velocity [$m \cdot sec^{-1}$]
α_c	Heat transfer coefficient (medium-side) [$W \cdot m^{-2} \cdot K^{-1}$]
α_k	Heat transfer coefficient (medium-side, includes metal wall resistance) [$W \cdot m^{-2} \cdot K^{-1}$]
δ	Product layer thickness [m]
δ_{max}	Maximum product layer thickness [mm]
δ_w	Metal wall thickness [m]
λ_s	Solid thermal conductivity [$W \cdot m^{-1} \cdot K^{-1}$]
λ_w	Metal wall thermal conductivity [$W \cdot m^{-1} \cdot K^{-1}$]
μ'	Dimensionless number ("sensible heat/latent heat" for "superheated" melt)
ρ_s	Solid specific mass [$kg \cdot m^{-3}$]
τ'	Dimensionless number ("time" for "superheated" melt)
ϕ_p	Belt cooler production [$kg \cdot sec^{-1}$]
ϕ_w	Cooling water flow [$kg \cdot sec^{-1}$]
χ	Dimensionless number ("layer thickness")

REFERENCES

Helber, H. D. (1993). *The Steel Belt Book*. Fellbach/Stuttgart, Germany: Sandvik Process Systems.

Preger, M. (1969). Drum flakers as heat exchangers. *Maschinenmarkt* 75:1622. In German.

Reith, T. (1976). Private communication.

Robens, A., Kaiser, M. (1990). Automatic pastillator for high capacities. *Verfahrenstechnik* 24:50. In German.

Sandvik (1980). New rapid cooling and shrink wrapping system for asphalt blocks. *Int. Pet. Times* 84:17.

Sandvik (1990). Belt cooling improves wax production. *Chem. Eng*. 97:145.

Schuster, D., Hodel, A. E. (1990). Rotary dropformer process gives once-through forming. *Chem. Process*. 53:107.

4

Drum Flakers

4.1. INTRODUCTION

The theory for the design of drum flakers was discussed in Chapter 2. Sec. 4.2 contains a general description of drum flakers. Attention is paid to the various ways in which the feed can be applied to the drum flaker: by pickup from a pan (standard method), by means of a bottom-mounted applicator roll, or by means of a top-mounted applicator roll. The successive steps of the drum flaking process are explained. The construction of drum flakers is discussed in Sec. 4.3. The relevance of small-scale tests is treated in Sec. 4.4. Secs. 4.5–4.8 deal with various aspects of a design method for drum flakers. Its application requires the knowledge of the physical properties of the flaked material. Verification of the design method is the subject of Sec. 4.9. A large amount of data for three different materials was analyzed.

A worked example of the design of a drum flaker is given in Sec. 4.10. It is also possible to scale-up the results of pilot plant trials and the procedure is discussed in Sec. 4.11.

Preger recommended a procedure to predict the effect of changing independent process variables of an existing drum flaker. Knowledge of physical properties is not required in the latter case. Preger's approach is explained in Sec. 4.12 and verified in Sec. 4.13.

4.2. GENERAL DESCRIPTION

Drum flakers are used to solidify melts continuously. A typical drum flaker is depicted in Fig. 4.1. The solidification of the melt is accomplished by applying a layer of product onto the outer surface of a hollow metal cylinder rotating around a horizontal axis while the inner surface is cooled by a coolant. The product is solidified, cooled, and removed by a doctor blade. Approximately 80% of the circumference of the cylinder is in contact with the product.

In many instances, flakes are obtained (thicknesses in the range 0.2–1.5 mm). Films, granules, needles, or powder can also be produced. The physical form of the product depends on the product characteristics, the process conditions, and the equipment settings. Broadly speaking, the products fresh from the drum flaker can be classified as brittle, elastic, or plastic. Most plastic or elastic products become brittle afterward due to postcrystallization. A good adhesion of the product to the flaker is essential. The adhesion is affected by the product characteristics (e.g., the change in specific mass of the solid on cooling), the properties of the material of construction of the flaker drum, and the prevailing temperatures. The presence of grooves in the surface of the drum can be helpful to promote adhesion. The material attaches itself readily to the material permanently present in the grooves. On the other hand, the contact can be too good and this leads to dust generation at the discharge point. Coating the drum with, e.g., Teflon, can provide a remedy.

Drum flakers can be classified according to the way the melt is applied onto the surface of the drum.

1. The drum dips to a certain depth into a pan filled with the molten process material. The metal cylinder picks up a film of process material. In many applications, the pan is shallow. In contrast, approximately 2/3 of the drum area is submerged when ice flakes are produced. Application by dipping

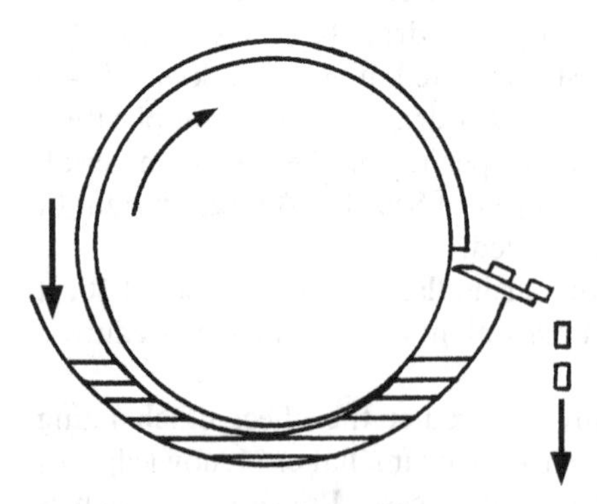

FIG. 4.1 A drum flaker with application from a pan.

is practiced for materials that adhere well to metal surfaces and crystallize readily. Application by dipping is a widely spread application method. One aspect is that product also grows on the lateral faces (see Fig. 4.1).

2. The melt is applied by a bottom-mounted applicator roll (see Fig. 4.2). The auxiliary roll dips into a shallow pan and the material picked up is transferred to the drum flaker. Now the lateral faces do not become covered with material. Furthermore, the size of the nip determines the thickness of the crystallized layer. The directions of rotation of the two cylinders can be equal or opposite.

3. The molten product is applied by means of a top-mounted applicator roll or from a trough. This method is used for materials that do not attach themselves properly to the metal surface and for materials that show post-crystallization. Double-drum flakers also fall into this category (see Figs. 4.3 and 4.4).

The discussion will now be focused on the drum flaker dipping into a shallow heated pan. However, many of the aspects to be discussed are equally applicable to other drum flaker types.

Regarding the crystallization process, four phases can be distinguished:

1. Wetting of the drum surface, followed by solidification of the melt on the surface of the drum. Wetting of the drum surface and solidification on the drum surface occurs while the drum surface part concerned is immersed in the melt.

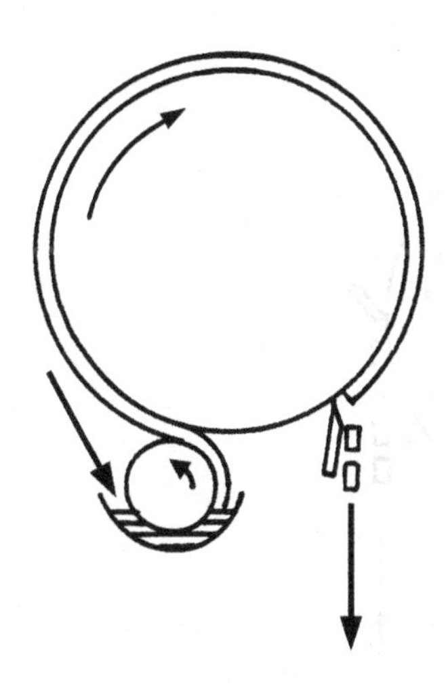

FIG. 4.2 A drum flaker with a bottom-mounted application roll.

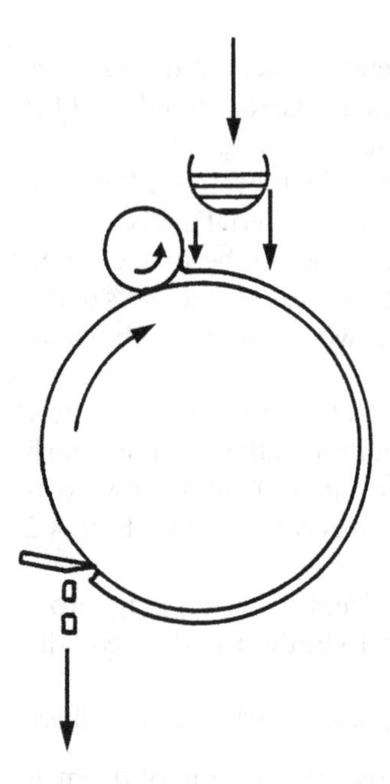

FIG. 4.3 A drum flaker with a top-mounted application roll.

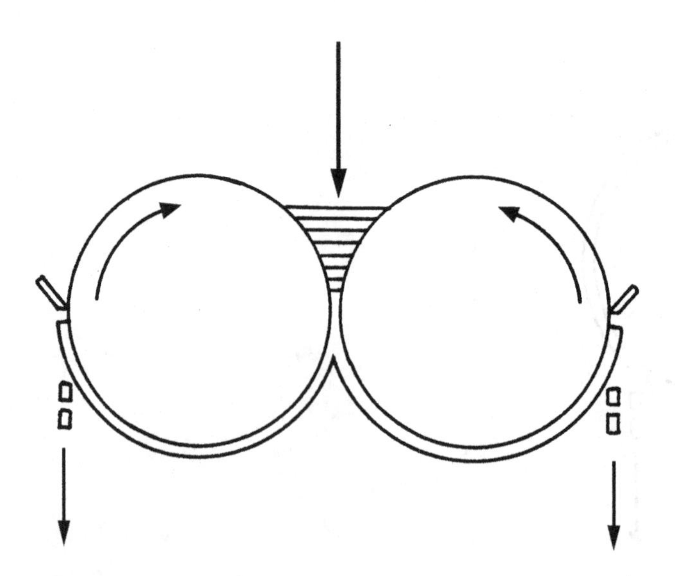

FIG. 4.4 A double-drum flaker.

2. A liquid layer is entrained when the drum emerges from the melt. This layer solidifies in the second phase. The cooling of the solid material from the first phase is continued.
3. The solidified material is cooled in the third phase. The flake discharge temperature is important since product caking characteristics on transfer and storage are directly related to it.
4. The solid layer removal by a doctor blade. This step entails the generation of dust if the material is brittle at this point. This is a drawback vis-à-vis pastillation and prilling.

The hourly production and the layer thickness are related as follows:

$$\phi_p = 3600 \cdot a w \rho_s L \delta \text{ kg} \cdot \text{hr}^{-1} \tag{4.1}$$

where a is an empirical coverage factor and is in the range 0.9–1.

A drum flaker is depicted in Fig. 4.5. The process of drum flaking is illustrated in Fig. 4.6. Fig. 4.7 gives an artist's view of a drum flaker. The

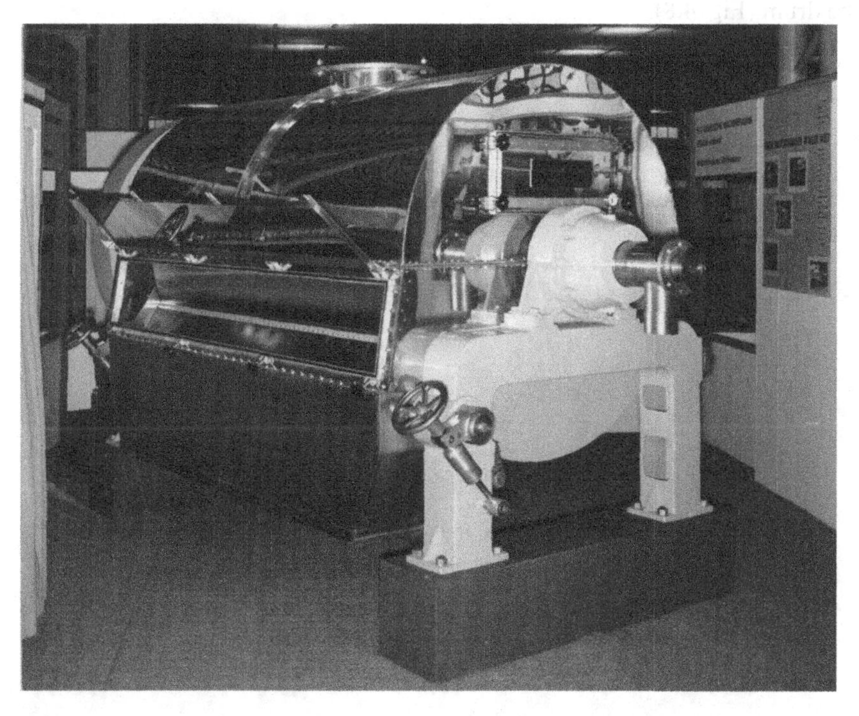

FIG. 4.5 A drum flaker at an exhibition. (Courtesy of GMF, Waddinxveen, The Netherlands.)

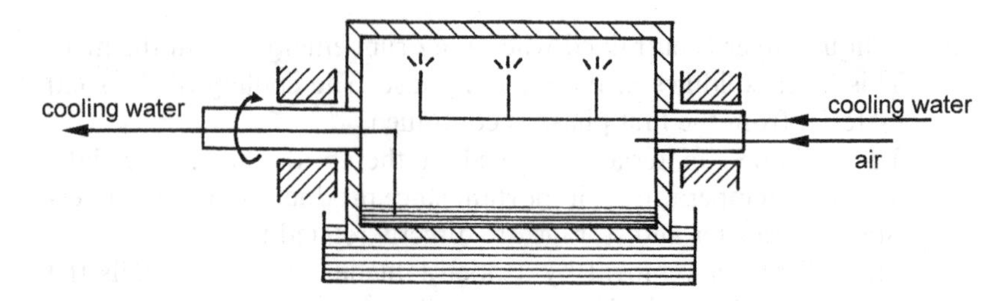

FIG. 4.6 The process of drum flaking.

cooling drum is mounted on hollow trunnions which are set in horizontal bearings. Usually, the coolant flows through the hollow shaft to manifolded nozzles from where the liquid is sprayed onto the inner drum surface. The liquid leaves the sump at the bottom through a dip pipe. The flow through this line is effected by air pressure. Alternatively, the cooling medium flows past the inner drum surface. A defined flow channel is provided in the inner part of the drum (Fig. 4.8).

FIG. 4.7 An artist's view of a drum flaker. (Courtesy of GMF, Waddinxveen, The Netherlands.)

The coolant flows from stationary parts into rotating parts and vice versa. This necessitates the installation of stuffing boxes or mechanical seals. The feed to the pan can be adjusted by a level control in the pan. Note that there is some degree of self-control: the layer thickness (and thus the production) increases when the level in the pan increases. The cylinder can be moved up or down hydraulically, pneumatically, or by hand.

The design of new drum flakers usually requires tests. It is possible to calculate the sizes of a drum flaker if the physical data of the material are known. An example will be treated in Sec. 4.10. However, it is advisable to check the adhesion of the processed material to the drum, the behavior during the crystallization process, and the scraping step. Furthermore, it is possible to verify the results of the calculations experimentally.

If the physical data are not known, the results of tests on a pilot drum flaker are scaled-up. The basis for scale-up is the layer thickness grown on the surface of the drum in a certain contact time. An example will be discussed in Sec. 4.11. A typical pilot drum flaker has a diameter of 0.5 m and a length of 0.5 m.

At the end of this section, some typical process parameters will be given:

- Melt temperature 60–160°C
- Rotational speed 1–20 rpm
- Melt viscosity 1–1000 mPa · s
- Layer thickness 0.2–1.5 mm
- Contact time (melt/drum surface) 1–60 sec
- Specific production capacity (reference: total drum cylindrical area) 100–500 kg · m^{-2} · hr^{-1}.

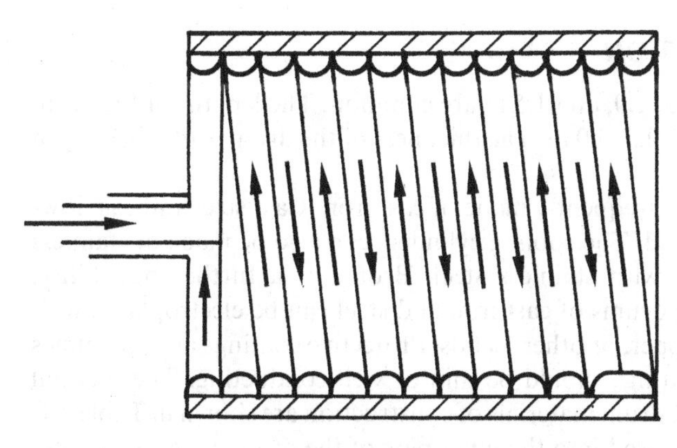

FIG. 4.8 Flow of a cooling medium through a defined flow channel.

Fig. 4.9 A closed drum flaker. (Courtesy of Robatel, Genas, France.)

4.3. CONSTRUCTION

Drum diameters of 0.5, 1.0, and 1.5 m are common. The lengths of the drum flakers are in the range 0.5–3.0 m. The thickness of the drum wall is usually in the range 10–20 mm.

Drum flakers are frequently made of cast iron. Cast steel and cast low-alloy steels are also used. The cooling cylinders can also be made of stainless steel or of steel clad with stainless steel. Bronze is a further possibility. Furthermore, cooling drums of cast iron and steel can be electroplated with chromium, nickel, copper, or other metals. Protective coatings are sometimes sprayed on; these coatings should be thin or well-conducting. The thermal conductivities of some drum materials of construction are shown in Table 4.1.

The coolant is passed into the inner part of the drum. One can choose from a set of coolants. Water is used most frequently (cooling tower water

TABLE 4.1 Thermal Conductivity of Materials of Construction for Drum Flakers at 20°C

Material of construction	λ_w ($W \cdot m^{-1} \cdot K^{-1}$)
Cast iron	42–63
Chromium	86
Copper (pure)	393
Nickel	58.5
Stainless steel 18% Cr, 8% Ni	21
Steel 0.2% C	50
Steel 0.6% C	46

Source: VDI Heat Atlas, 1993.

25–30°C, surface water 20°C, well water 15°C, and chilled water 5°C). The amount of water fed is such that the temperature increases by a few degrees only. It is also possible to use brine; temperatures as low as −35°C are then possible. Flashing coolants like ammonia and propylene can also be used. The proper distribution of the coolant is essential for a uniform flake thickness. The feasibility of the coolant should be checked. For example, the adhesion of the solid layer to the metal surface can be impacted by the temperature of the coolant. Furthermore, it is possible that the crystallization is retarded by low coolant temperatures.

With belt coolers, crosscontamination product/coolant is improbable. Crosscontamination is impossible when working with drum flakers.

A closed construction can be applied if the material processed can be readily oxidized by the oxygen in the air, is hygroscopic, is acted upon by carbon dioxide, or is sensitive to dust in the atmosphere. Furthermore, closed constructions are applied for materials that emit (toxic) vapors, sublime, evaporate, or are toxic. Figure 4.9 shows a closed construction.

4.4. SMALL-SCALE INVESTIGATIONS

The feasibility of the solidification of a melt on a metal surface can be checked by pouring some of the liquid material on a metal plate. The plate is cooled by means of, for example, cooling water. The times for solidification and cooling can be measured and these times can be compared to the times which can be calculated by applying the relationships derived in Chapter 2 (see Sec. 3.8.1). The relevant physical properties must be known for such a comparison. Many of the observations made in Sec. 3.3 are also applicable for drum flakers. See, for example, the remarks regarding the crystallinity of the material, the impact of the cooling rate, and the adhesion of the material to the metal.

It is often useful to carry out investigations on a laboratory drum flaker. The diameter of the roll and the length of the roll can be, for example, 0.2 m. The impact of the independent process variables can be checked quickly. However, scale-up work should be carried out on a drum flaker having a roll diameter and a roll length of, for example, 0.5 m. The independent process variables are:

- The rotational speed of the drum
- The "wetted length" of the drum
- The feed temperature
- The cooling water temperature.

The process results are also impacted by the material(s) of construction and the wall thickness of the drum flaker. Furthermore, the way the cooling water is applied plays a role.

The scale-up criterion is the contact time. This means that, for equal contact times, it is assumed that equal layer thicknesses are obtained on the two scales. Of course, the other independent process variables must be kept constant. This statement is approximately true, especially if the entrainment effects are comparable.

The execution of scale-up experiments will be discussed in Sec. 4.11. For successful scale-up of the results of pilot plant trials, it is not necessary to know the full set of physical data of the material processed.

4.5. CONVECTIVE HEAT TRANSFER FROM THE MELT TO THE SOLIDIFIED LAYER

The recommended correlation is (VDI-Heat Atlas, 1993)

$$Nu_{\mathrm{lam}} = 0.664\sqrt{Re}\sqrt[3]{Pr} \tag{4.2}$$

$$Re = \frac{wl}{\nu} < 10^5$$

$$Pr = \frac{\mu c_l}{\lambda_l} \quad (0.6 < Pr < 2000)$$

$$Nu_{\mathrm{lam}} = \frac{\alpha_o l}{\lambda_l}$$

There is also a correlation for turbulent flow; however, that correlation is valid for Reynolds numbers in the range $5 \cdot 10^5$–10^7. It is therefore recommended to use the aforementioned correlation for Reynolds numbers smaller than $5 \cdot 10^5$.

The relation is valid for flat plates over which a fluid flows. l is the length of the plate in the direction of the flow. The physical properties are assessed at T_f, the temperature of the feed.

For drum flakers, l is taken as the "wetted length" of the drum flaker.

Usually, the temperature of the feed is only 5–10 K higher than the melting point. The effect of the convective heat transfer from the melt to the solidified layer is not very important if this small temperature difference applies. The correlation still provides the possibility to approach reality.

4.6. SOLIDIFICATION IN THE TROUGH

The solidification in the trough occurs on a wall with a variable temperature. Basically, two different approaches are possible. The first possible approach is the one described as Case 1 in Sec. 2.5. Here the melt is "superheated" (i.e., it has a temperature higher than its melting point), well-mixed, and in contact with the growing layer. There is heat transfer to the growing layer due to the crystallization and due to convective heat transfer. The design relations are summarized as follows.

$$\theta = \frac{T_f - T_o}{T_o - T_c} \cdot \frac{\alpha_o}{\alpha_k}$$

$$\chi = \frac{\alpha_k \delta}{\lambda_s}$$

$$\mu = \frac{c_s(T_o - T_c)}{2i}$$

$$F_0 = -\frac{1}{\theta^2}\ {}^{e}\log\left(1 - \frac{\theta\chi}{1-\theta}\right) - \frac{\chi}{\theta}$$

$$F_1 = -\left(\frac{1}{\theta_2} - 1\right) {}^{e}\log\left(1 - \frac{\theta\chi}{1-\theta}\right) - \frac{\chi}{\theta} - {}^{e}\log(\chi + 1)$$

$$\tau = F_0 + \mu F_1$$

$$t = \frac{\tau i \rho_s \lambda_s}{\alpha_k^2 (T_o - T_c)} \text{ sec}$$

The second possible approach is the one described as Case 3 in Sec. 2.5. In this case, the melt is also "superheated." However, the melt is assumed to be stagnant and nonconductive. Then both the "superheat" and the heat of crystallization are transferred when the material solidifies. This approach is applied for the design of belt coolers. The design relations are summarized as follows.

$$\chi = \frac{\alpha_k \delta}{\lambda_s}$$

$$\mu' = \frac{c_s(T_o - T_c)}{2\{i + c_l(T_f - T_o)\}}$$

$$\tau' = \chi + \frac{\chi^2}{2} + \mu'\left\{\chi + \frac{\chi^2}{2} - {}^{e}\log(\chi + 1)\right\} \tag{2.17}$$

$$t = \frac{\tau'\{i + c_l(T_f - T_o)\}\rho_s\lambda_s}{\alpha_k^2(T_o - T_c)} \text{ sec}$$

It was shown in Sec. 2.5 that the results obtained by applying the two approaches are not very different when the "superheat" is approximately 10 K. A further aspect is that the application of the method for belt coolers is intrinsically simpler than the application of the method according to Case 1 of Sec. 2.5. The main reason is that, first, for the latter method, a heat transfer coefficient for the convective heat transfer from the melt to the growing layer must be assumed. By means of an iterative process, the assumption must be confirmed.

However, this so-called complete design method for the solidification process takes the successive steps into consideration. The discussion of the pros and cons of the two design methods will be continued in Sec. 4.10. In both cases, a simplification is implicitly introduced. The enthalpy change of the drum material is neglected (the specific heat of the metal is assumed to be zero). This is a conservative approach as, in actual fact, heat is stored in the metal wall while the process material crystallizes.

4.7. MELT ENTRAINMENT AND SOLIDIFICATION OF THE ENTRAINED MELT

Liquid product is entrained from the trough by the rotating drum. Groot Wassink (1977) gives a correlation to calculate the entrainment:

$$\frac{q}{\nu} = \frac{1}{2}\left(\frac{w^3}{g\nu}\right)^{0.5} \tag{4.3}$$

where q is the entrainment in $m^3 \cdot m^{-1} \cdot sec^{-1}$. q is a function of the drum speed w in $m \cdot sec^{-1}$ and the kinematic viscosity ν in $m^2 \cdot sec^{-1}$. The kinematic viscosity is taken at a temperature which is the average of the melt temperature and the melting point. q/w is the thickness of the entrained layer in m. Groot Wassink reported the measurements leading to the correlation in 1976. Experiments were carried out on a pilot-plant drum flaker and a hydrodynamic model of a drum flaker. The data of these test series are as follows.

Pilot-plant drum flaker
Diameter: 0.5 m
Drum length: 0.5 m

Test liquids: water and glycerol/water mixtures
w-values in the range 0.112–1.12 $\mathrm{m \cdot sec^{-1}}$
ν-values in the range $1.15 \cdot 10^{-6}$–$59.0 \cdot 10^{-6}\ \mathrm{m^2 \cdot sec^{-1}}$

Hydrodynamic model of a drum flaker
Diameter: 0.197 m
Length: 0.5 m
Test liquid: water
w-values in the range 0.399–1.73 $\mathrm{m \cdot sec^{-1}}$
$\nu = 10^{-6}\ \mathrm{m^2 \cdot sec^{-1}}$

The solidification of the entrained melt is treated according to the approach for belt coolers. This will be shown in Sec. 4.10.

The entrainment can be neglected if the drum circumferential velocity is low ($<0.1\ \mathrm{m \cdot sec^{-1}}$) and the melt viscosity is only several millipascal seconds. Neglecting the entrainment simplifies the design procedure for a drum flaker markedly.

4.8. COOLING OF THE SOLIDIFIED LAYER

A simple approach is suggested.

$$Q_c = \phi_p c_s (T_o - \bar{T})\ \mathrm{W}$$

This is the heat load for cooling. It is assumed that the layer is initially at the melting point.

$$Q_c = k_c F \Delta T_m\ \mathrm{W}$$

This is the heat load that can be carried away. The heat load that can be carried away is assumed to be equal to the heat load that should be carried away.

The heat transfer coefficient k_c is defined as follows:

$$\frac{1}{k_c} = \frac{1}{\alpha_c} + \frac{\delta_w}{\lambda_w} + \frac{\delta}{\lambda_s}\ \mathrm{W^{-1} \cdot m^2 \cdot K} \tag{4.4}$$

It follows from Eq. (4.4) that it is assumed that the heat from the layer has to travel through the layer in the first place.

A further assumption is that the area for cooling is 50% of the drum cylindrical area.

$$\Delta T_m = \frac{(T_o - T_c) - (\bar{T} - T_c)}{{}^{e}\log \dfrac{(T_o - T_c)}{(\bar{T} - T_c)}}\ \mathrm{K}$$

It follows that:

$$\phi_p c_s(T_o - \overline{T}) = k_c F \cdot \frac{(T_o - T_c) - (\overline{T} - T_c)}{{}^e\log \dfrac{T_o - T_c}{\overline{T} - T_c}}$$

and

$$\overline{T} = T_c + \frac{T_o - T_c}{\exp \dfrac{k_c F}{\phi_p c_s}} \ ^\circ\mathrm{C}$$

This simple model also neglects the enthalpy change of the metal wall. In this case, this assumption is not conservative as heat must be extracted from the metal wall by the cooling medium.

4.9. VERIFICATION OF THE DESIGN METHODS

4.9.1. Small-Scale Trials with Caprolactam

Physical Properties

$T_o = 69.4°\mathrm{C}$
$i = 142{,}491\ \mathrm{J \cdot kg^{-1}}$
$\rho_l = 1020\ \mathrm{kg \cdot m^{-3}}$ (77.15°C)
$\rho_s = 1096\ \mathrm{kg \cdot m^{-3}}$
$c_l = 2160\ \mathrm{J \cdot kg^{-1} \cdot K^{-1}}$ (77°C)
$c_s = 1526\ \mathrm{J \cdot kg^{-1} \cdot K^{-1}}$ (44.5°C)
$\lambda_l = 0.1589\ \mathrm{W \cdot m^{-1} \cdot K^{-1}}$ (89.36°C)
$\lambda_s = 0.230\ \mathrm{W \cdot m^{-1} \cdot K^{-1}}$ (38°C) (Groot Wassink, 1976)
$\mu = 1.97 \cdot 10^{-2}\ \mathrm{N \cdot sec \cdot m^{-2}}$ at 70.15°C
$\mu = 9.0 \cdot 10^{-3}\ \mathrm{N \cdot sec \cdot m^{-2}}$ at 78.15°C

Equipment

$L = 0.5\ \mathrm{m}$
$D = 0.5\ \mathrm{m}$
$\delta_w = 9\ \mathrm{mm}$
$\lambda_w = 15\ \mathrm{W \cdot m^{-1} \cdot K^{-1}}$ (stainless steel, 20°C)

The cooling water was sprayed.

α_k

$$\frac{1}{\alpha_k} = \frac{1}{\alpha_c} + \frac{\delta_w}{\lambda_w}\ \mathrm{W^{-1} \cdot m^2 \cdot K}$$

α_c is in the range 1000–3000 kcal · m^{-2} · hr^{-1} · K^{-1} (Preger, 1969). Hence α_k is in the range 685–1128 W · m^{-2} · K^{-1}. For the calculations, $\alpha_k = 900$ W · m^{-2} · K^{-1}.

General Description

Four different series of trials comprised 33 experiments. The test results were made available by GMF at Waddinxveen in The Netherlands. The caprolactam qualities used were technical qualities and came from different sources. Three experiments had to be excluded from the interpretations because the sets of data were incomplete. The material balances of the 30 remaining experiments were checked. This was done by comparing the measured production with the calculated production. The calculated production was obtained by means of Eq. (4.1). One measurement was excluded from the interpretations because the difference between the calculated production and the measured production was greater than 15%. It appeared that a is approximately 0.95. The data of the ultimate 29 experiments are listed in Appendix A.

Ranges of independent variables (29 experiments):

n: 1–10 min^{-1}
h: 10–140 mm
T_f: 76.5–95°C
T_c: 0–34°C
t_e: 1–9 min

Ranges of dependent variables:

ϕ_p: 50–300 kg · hr^{-1} (65–380 kg · m^{-2} · hr^{-1})
δ_{act}: 0.42–1.1 mm
t_{act}: 0.5–17.6 sec
T_{act}: 9–67.5°C

Description of the Crystallization

This description covers both the crystallization in the pan and the crystallization of the entrained liquid layer. It was checked whether the thickness of the crystallized layers can be predicted. Two fundamentally different mechanisms are active:

- The growth of a layer in the pan
- The crystallization of an entrained liquid layer.

The complete model outlined in Sec. 4.6 was employed to predict the thickness of the layer grown in the pan. The approach described in Sec. 4.7 was used to predict the thickness of the entrained liquid layer. The two predicted layer thicknesses were added and compared to the layer thickness actually found. The predictions were on the average 10% greater than the actual layer thicknesses. In other words, $\delta_{pred}/\delta_{act} = 1.10$ (standard deviation 0.10). Two measurements were excluded because the results were flyers (Nos. 1 and 16). So 27 measurements were taken into account.

It is possible that the predictions would match the experimental findings even better if more information regarding α_k were available.

There is a tendency that $\delta_{pred}/\delta_{act}$ is in the range 1–1.1 for low rotational speeds (1, 2, or 3 min^{-1}) and in the range 1.1–1.2 for high rotational speeds (up to 10 min^{-1}).

It is interesting to note that the predicted thickness of the entrained layer is about equal to the predicted thickness of the layer grown in the pan at high rotational speed. For example, at 10 min^{-1}, both layers have a thickness of approximately 0.3 mm (contact time 2 sec). The strong influence of the entrainment mechanism is caused by caprolactam's high viscosity; that is, at 78.15°C, the melt is nine times as viscous as water at 20°C.

The time required for the solidification of the entrained liquid layer can be calculated by applying the model for belt coolers twice. The time required for the crystallization of the full layer is calculated and the time for the crystallization of the layer grown in the pan. The two times are subtracted. This will be shown in Sec. 4.10. However, experimental verification was not possible. It was not necessary either as the area for cooling is assumed to be equal to 50% of the drum area.

Description of the Cooling

The product exit temperature could be rather accurately predicted by applying the model described in Sec. 4.8. The actual temperature drop ($T_o - T_{act}$) was compared to the calculated temperature drop ($T_o - \overline{T}$).

The predicted temperature drop was on the average 8% greater than the actual temperature drop. In other words, the ratio was 1.08 (standard deviation 0.09). Three measurements were excluded because the results were flyers (Nos. 1, 15, and 17). So 26 measurements were considered. Again, it is possible that the predictions would match the experimental findings even better if there would be more information regarding α_k.

4.9.2. Small-Scale Trials with Naphthalene

Physical Properties

$T_o = 80.4°C$
$i = 148{,}100\ J \cdot kg^{-1}$
$\rho_l = 975\ kg \cdot m^{-3}$ (82°C)

$\rho_s = 1150 \text{ kg} \cdot \text{m}^{-3}$ (60°C)
$c_l = 1710 \text{ J} \cdot \text{kg}^{-1} \cdot \text{K}^{-1}$ (85°C)
$c_s = 1440 \text{ J} \cdot \text{kg}^{-1} \cdot \text{K}^{-1}$ (51°C)
$\lambda_l = 0.135 \text{ W} \cdot \text{m}^{-1} \cdot \text{K}^{-1}$ (89°C)
$\lambda_s = 0.37 \text{ W} \cdot \text{m}^{-1} \cdot \text{K}^{-1}$ (38°C) (Groot Wassink, 1976)
$\mu = 0.96 \cdot 10^{-3} \text{ N} \cdot \text{sec} \cdot \text{m}^{-2}$ (80.4°C)
$\mu = 0.846 \cdot 10^{-3} \text{ N} \cdot \text{sec} \cdot \text{m}^{-2}$ (90.15°C)

Equipment

$L = 0.5$ m
$D = 0.5$ m
$\delta_w = 8$ mm
$\lambda_w = 15 \text{ W} \cdot \text{m}^{-1} \cdot \text{K}^{-1}$ (stainless steel, 20°C)

The cooling water was sprayed.

α_k

$$\frac{1}{\alpha_k} = \frac{1}{\alpha_c} + \frac{\delta_w}{\lambda_w} \quad \text{W}^{-1} \cdot \text{m}^2 \cdot \text{K}$$

α_c is in the range 1000–3000 $\text{kcal} \cdot \text{m}^{-2} \cdot \text{hr}^{-1} \cdot \text{K}^{-1}$ (Preger, 1969). Hence α_k is in the range 718–1220 $\text{W} \cdot \text{m}^{-2} \cdot \text{K}^{-1}$.

For the calculations, $\alpha_k = 1000 \text{ W} \cdot \text{m}^{-2} \cdot \text{K}^{-1}$.

General Description

A series of nine experiments was carried out. The test results were made available by GMF at Waddinxveen in The Netherlands. The naphthalene used was a technical quality. One trial (No. 4) had to be excluded from the interpretations because only powder was produced (a high drum rotational speed, that is, 13.3 min^{-1}, was combined with an immersion depth of 5 mm only). Table 4.2 contains the independent process variables of the tests, while Table 4.3 summarizes the dependent process variables.

The material balances of the remaining eight experiments were checked. This was done by comparing the measured production with the calculated production. The calculated production was obtained by means of Eq. (4.1). The differences between the calculated production and the measured production were smaller than 20% in all cases and smaller than 10% in seven cases. It appeared that a is approximately 1.05.

Description of the Crystallization

This description covers both the crystallization in the pan and the crystallization of the entrained liquid layer.

TABLE 4.2 Independent Process Variables of Nine Naphthalene Drum Flaking Trials

Trial No.	1	2	3	4	5	6	7	8	9
n (min^{-1})	16	16.2	16.3	13.3	3.5	5	3	2	3.6
h (mm)	30	10	5	5	90	100	100	100	100
T_f (°C)	82.0	82.4	82.0	82.5	92.0	92.0	84.0	86.0	87.0
T_c (°C)	34.3	34.5	29.1	25.7	28.8	29.2	31.2	32.3	33.1
t_e (min)	2	2	2	2	2	2	2	2	2

Also in this case, it was checked whether the thickness of the crystallized layers could be predicted. The approach chosen was identical to the approach chosen for caprolactam. Three more measurements were excluded because the predictions were much smaller than the actual layer thicknesses. These experiments were Nos. 1, 2, and 3. The contact times were substantially smaller than 1 sec. The contact times of Experiments 5, 6, 7, 8, and 9 were in the range 3.5–8.8 sec. It appears that the layer thicknesses cannot be predicted for contact times shorter than 1–1.5 sec. A tentative explanation is that the cold drum attracts a crystallized layer almost instantaneously on contacting the melt. However short the contact time, the minimum layer thickness is 0.25–0.3 mm for naphthalene.

The predictions were on the average 10% smaller than the actual layer thicknesses. In other words, $\delta_{pred}/\delta_{act} = 0.90$ (range 0.85–0.95). It is possible that the predictions would match the experimental findings even better if there would be more information regarding α_k.

The inability to predict the layer thickness for very short contact times was not found for caprolactam. This is probably caused by caprolactam's relatively high viscosity. Very short contact times occur when the rotational speed is high. A relatively thick caprolactam layer is entrained at high rota-

TABLE 4.3 Dependent Process Variables of Nine Naphthalene Drum Flaking Trials

Trial No.	1	2	3	4	5	6	7	8	9
δ_{act} (mm)	0.4	0.35	0.35	[a]	0.70	0.60	0.85	1.0	0.8
T_{act} (°C)	46.8	44.5	41.9	33.8	36.1	41.4	35.7	38.5	41.4
ϕ_p (kg · hr^{-1})	289.2	270.9	278.3	205.8	139.8	170.3	135	108	151.5
l (m)	0.25	0.14	0.09	0.09	0.44	0.46	0.46	0.46	0.46
t_{act} (sec)	0.60	0.33	0.21	0.26	4.89	3.5	5.8	8.8	4.89

[a] Only powder was produced.

tional speed. That is, an entrainment of approximately 0.3 mm is predicted for a rotational speed of 10 min^{-1}. For naphthalene, an entrainment of approximately 0.1 mm is predicted for a rotational speed of 16 min^{-1}. The point made will be illustrated with an approximate survey.

	Naphthalene	Caprolactam
Rotational speed, min^{-1}	16	10
t_{act}, sec	0.5	1
δ_{act}, mm	0.4	0.4
Predictions		
δ_{ent}, mm	0.1	0.3
δ_{pan}, mm	0.1 +	0.1 +
	0.2	0.4

Summing up, the overall conclusion is that the predictions of the layer thickness are unreliable for contact times shorter than 1–1.5 sec. However, this effect can be masked for viscous melts by entrainment.

The calculation of the time required for the crystallization of the entrained liquid layer was discussed in the previous section on caprolactam.

Description of the Cooling

The product exit temperature could be rather accurately predicted by applying the model described in Sec. 4.8. The actual temperature drop ($T_o - T_{act}$) was compared to the calculated temperature drop ($T_o - \overline{T}$). Eight measurements were considered (Nos. 1, 2, 3, 5, 6, 7, 8, and 9). The predicted temperature drop was on the average 7% greater than the actual temperature drop. In other words, the ratio was 1.07 (range 0.97–1.16).

Concluding Remarks

Again, it is possible that the predictions would match the experimental findings even better if there were more information regarding α_k.

It is apparent that the model can describe the cooling results of all eight trials.

The models for the prediction of the thickness of the crystallized layers cannot describe the results of trials with a very short contact time (shorter than 1.5 sec).

4.9.3. Large-Scale Trials with a Product (Kloppenburg, 1968)

Physical Properties

$T_o = 61\,°C$
$i = 138{,}000\ J \cdot kg^{-1}$
$\rho_l = 1355\ kg \cdot m^{-3}$ (80°C)
$\rho_s = 1360\ kg \cdot m^{-3}$ (25°C)
$c_l = 1880\ J \cdot kg^{-1} \cdot K^{-1}$ (80°C)
$c_s = 1610\ J \cdot kg^{-1} \cdot K^{-1}$ (25°C)
$\lambda_l = 0.30\ W \cdot m^{-1} \cdot K^{-1}$ (80°C)
$\lambda_s = 0.26\ W \cdot m^{-1} \cdot K^{-1}$ (25°C)
$\mu = 2.07 \cdot 10^{-3}\ N \cdot sec \cdot m^{-2}$ (70°C)
$\mu = 1.76 \cdot 10^{-3}\ N \cdot sec \cdot m^{-2}$ (80°C)

Equipment

First series of trials:

$D = 1$ m

The drum was filled with cooling water flowing through the drum.

$\alpha_k = 500\ W \cdot m^{-2} \cdot K^{-1}$ (assumption)

Second series of trials:

$D = 1.5$ m

The drum was filled with cooling water flowing through the drum.

$\alpha_k = 500\ W \cdot m^{-2} \cdot K^{-1}$ (assumption)

General Description

The first series of trials comprised five experiments. The rotational speed of the drum was varied while the contact angle was kept constant. The length of the drum was probably 3 m.

The second series also comprised five experiments. The contact angle was varied while the rotational speed was fixed. The length of the drum was probably 3 m. Independent and dependent variables: see Tables 4.4 and 4.5.

Description of the Crystallization

This description covers both the crystallization in the pan and the crystallization of the entrained liquid layer.

TABLE 4.4 Independent and Dependent Variables of the First Series of Trials Concerning the Large-Scale Flaking of a Product

Test No.	1	2	3	4	5
n (min^{-1})	1.34	1.83	2.40	3.26	4.38
δ_{act} (mm)	0.67	0.62	0.49	0.445	0.36
ϕ_p (kg · hr^{-1})	725	800	880	1045	1170
t_{act} (sec)	11.3	8.3	6.3	4.7	3.5

$h = 0.15$ m.
$T_f = 77.5$°C.
$T_c = 16$°C.

Again, it was checked whether the thickness of the crystallized layers could be predicted. The approach chosen was identical to the approaches chosen for caprolactam and naphthalene. For both series of trials, the predictions were almost identical to the actual layer thicknesses. One aspect is that the minimum contact time was 3.5 sec. Hence deviations for very short contact times were not found (see Sec. 4.9.2 regarding naphthalene experiments). A further interesting aspect is that the viscosity of the product is in between the viscosities of naphthalene and caprolactam. With caprolactam, the viscosity causes a substantial entrainment at high rotational speeds. For naphthalene, the entrainment can almost be neglected. The calculated thickness of the entrained layer of the product is in the range 10–20% of the total layer thickness. For equal contact time, the layer thickness of the second series of trials is smaller than the layer thickness of the first series of experiments (see Tables 4.4 and 4.5). The reason is the greater entrainment at the first series of trials because of the greater circumferential speed. The assumption of $\alpha_k = 500$ W · m^{-2} · K^{-1} is a further aspect. α_k includes the

TABLE 4.5 Independent and Dependent Variables of the Second Series of Trials Concerning the Large-Scale Flaking of a Product

Test No.	1	2	3	4	5
h (m)	0.034	0.046	0.067	0.087	0.111
δ_{act} (mm)	0.38	0.42	0.47	0.52	0.56
ϕ_p (kg · hr^{-1})	570	615	680	730	780
t_{act} (sec)	4.7	5.4	6.5	7.4	8.4

$n = 1.25$ min^{-1}.
$T_f = 77.5$°C.
$T_c = 16$°C.

resistances to heat transfer from the cooling water to the metal wall and through the metal wall. The cooling water flowed along the heat exchanging area. Of course, the results of the calculations depend on α_k.

Almost perfect agreement between theory and practice for both series of trials was obtained on taking $\alpha_k = 500\ \mathrm{W \cdot m^{-2} \cdot K^{-1}}$.

Description of the Cooling

Data regarding the cooling are not available.

Final Remarks

This product was the same product as used for the verification of the model for belt coolers. It was shown that the model for belt coolers was applicable for both small-scale tests and large-scale trials (see Sec. 3.8). For those

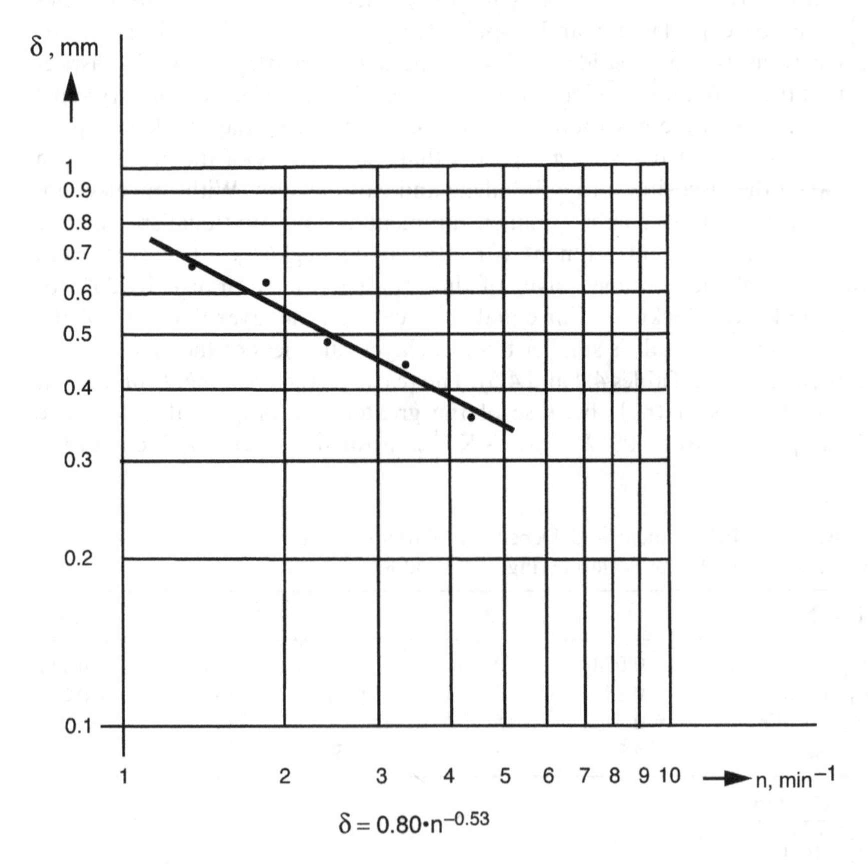

Fig. 4.10 A Preger plot (δ).

verifications, there was more certainty regarding α_k because the cooling water was sprayed and belt data were available. Moreover, the exact α_k-value is less important when the flake thickness is greater than, e.g., 2 mm. In that case, the resistance to heat transfer of the crystallized layer is much greater than $1/\alpha_k$.

The measurements of the first series of trials were also interpreted according to the "quadratic growth law" (see Sec. 2.3). $\delta_{act} = 0.80 \cdot n^{-0.53}$ mm describes the experimental results approximately (see Sec. 4.13 and Fig. 4.10).

4.10. DESIGN OF A DRUM FLAKER

It is required to design a drum flaker for the production of 1 $t \cdot hr^{-1}$ naphthalene flakes having a thickness of 1 mm. The diameter of the drum flaker is taken as 1.5 m while the wall thickness of the flaking drum is 10 mm. The immersion depth is 300 mm. The flaking drum is made of stainless steel AISI 316. The average cooling water temperature is 32°C. The liquid in the pan has a temperature of 86°C. The average flake temperature should be in the range 30–50°C.

Physical data of naphthalene:
See Sec. 4.9.2.

Design

First, the contact time for a 0.95-mm layer is calculated.
$\alpha_o = 70\ W \cdot m^{-2} \cdot K^{-1}$ (assumption, to be checked)
$\alpha_c = 3000\ W \cdot m^{-2} \cdot K^{-1}$ (assumption for a water spray)
$\lambda_w = 15\ W \cdot m^{-1} \cdot K^{-1}$

$$\frac{1}{\alpha_k} = \frac{1}{3000} + \frac{0.01}{15} \rightarrow \alpha_k = 1000\ W \cdot m^{-2} \cdot K^{-1}$$

$$\theta = \frac{T_f - T_o}{T_o - T_c} \cdot \frac{\alpha_o}{\alpha_k} = \frac{86 - 80.4}{80.4 - 32} \cdot \frac{70}{1000} = 0.0081$$

$$\chi = \frac{\alpha_k \delta}{\lambda_s} = \frac{1000 \cdot 0.00095}{0.37} = 2.57$$

$$\mu = \frac{c_s(T_o - T_c)}{2i} = \frac{1440(80.4 - 32)}{2 \cdot 148,100} = 0.235$$

$$F_0 = -\frac{1}{\theta^2}\,{}^e\!\log\left(1 - \frac{\theta\chi}{1-\theta}\right) - \frac{\chi}{\theta} = 323.3 - 317.3 = 6.0$$

$$F_1 = -\left(\frac{1}{\theta^2} - 1\right){}^e\!\log\left(1 - \frac{\theta\chi}{1-\theta}\right) - \frac{\chi}{\theta} - {}^e\!\log(\chi + 1)$$

$$= 323.3 - 317.3 - 1.3 = 4.7$$

$$\tau = F_0 + \mu F_1 = 6.0 + 0.235 \cdot 4.7 = 7.1$$

$$t = \frac{\tau i \rho_s \lambda_s}{\alpha_k^2 (T_o - T_c)} = \frac{7.1 \cdot 148,100 \cdot 1150 \cdot 0.37}{1000^2 (80.4 - 32)} = 9.2 \text{ sec}$$

The contact length is 1.39 m (corresponds with an immersion depth of 300 mm).

A full rotation is completed in 31.2 sec (1.92 rpm). $w = 0.151\ \text{m} \cdot \text{sec}^{-1}$

Calculation of α_o:

$$Re = \frac{wl}{\nu} = \frac{0.151 \cdot 1.39 \cdot 975}{0.9 \cdot 10^{-3}} = 227,381$$

$$Pr = \frac{\mu c_l}{\lambda_l} = \frac{0.9 \cdot 10^{-3} \cdot 1710}{0.135} = 11.4$$

$$Nu_{\text{lam}} = 0.664 \sqrt{Re} \sqrt[3]{Pr} = 0.664 \cdot 477 \cdot 2.25 = 713$$

$$\alpha_o = \frac{Nu_{\text{lam}} \cdot \lambda_l}{l} = \frac{713 \cdot 0.135}{1.39} = 69.2\ \text{W} \cdot \text{m}^{-2} \cdot \text{K}^{-1}$$

This value is close to the assumption $\alpha_o = 70\ \text{W} \cdot \text{m}^{-2} \cdot \text{K}^{-1}$.

Melt entrainment:

$$\nu = \frac{0.9 \cdot 10^{-3}}{975} = 9.2 \cdot 10^{-7}\ \text{m}^2 \cdot \text{sec}^{-1}$$

$$w = 0.151\ \text{m} \cdot \text{sec}^{-1}$$

$$q = \nu \cdot \frac{1}{2} \left(\frac{w^3}{g \cdot \nu} \right)^{0.5} = 9.2 \cdot 10^{-7} \cdot \frac{1}{2} \left(\frac{0.151^3}{9.81 \cdot 9.2 \cdot 10^{-7}} \right)^{0.5}$$

$$= 8.98 \cdot 10^{-6}\ \text{m}^3 \cdot \text{m}^{-1} \cdot \text{sec}^{-1}$$

$$\frac{q}{w} = \frac{8.98 \cdot 10^{-6}}{0.151} = 5.9 \cdot 10^{-5}\ \text{m}$$

The thickness of the entrained layer is 0.06 mm.

Material balance:

$$\phi_p = 3600 \cdot a w \rho_s L \delta\ \text{kg} \cdot \text{hr}^{-1}$$

$$L = \frac{1000}{3600 \cdot 0.9 \cdot 0.151 \cdot 1150 \cdot 1.01 \cdot 10^{-3}} = 1.76\ \text{m}$$

The length of the drum flaker is 1.80 m and the specific production rate is 118 $\text{kg} \cdot \text{m}^{-2} \cdot \text{hr}^{-1}$.

Solidification time for the entrained melt

The design method for belt coolers is used. First, the time for the solidification of a layer of 1.01 mm is calculated. Next, the time for the crystallization of a layer having a thickness of 0.95 mm is calculated. The latter time is subtracted from the former time.

$$\chi = \frac{\alpha_k \delta}{\lambda_s} = \frac{1000 \cdot 1.01 \cdot 10^{-3}}{0.37} = 2.73 \ (1.01 \text{ mm})$$

$$\mu' = \frac{c_s(T_o - T_c)}{2\{i + c_l(T_f - T_o)\}} = \frac{1440(80.4 - 32)}{2\{148,100 + 1710(86 - 80.4)\}} = 0.221$$

$$\tau' = \chi + \frac{\chi^2}{2} + \mu'\left\{\chi + \frac{\chi^2}{2} - {}^e\log(\chi + 1)\right\}$$

$$= 2.73 + \frac{2.73^2}{2} + 0.221\left\{2.73 + \frac{2.73^2}{2} - {}^e\log(2.73 + 1)\right\} = 7.59$$

$$t = \frac{\tau'\{i + c_l(T_f - T_o)\}\rho_s \lambda_s}{\alpha_k^2 (T_o - T_c)}$$

$$= \frac{7.59\{148,100 + 1710(86 - 80.4)\}1150 \cdot 0.37}{1000^2(80.4 - 32)} = 10.52 \text{ sec}$$

Calculation for 0.95 mm:

$$\chi = \frac{1000 \cdot 0.95 \cdot 10^{-3}}{0.37} = 2.57$$

$$\mu' = 0.221$$

$$\tau' = 6.89$$

$$t = 9.55 \text{ sec}$$

The solidification of the entrained melt takes 10.52–9.55 = 0.97 sec (11.2°).

Cooling of the flakes:

$$\frac{1}{k_c} = \frac{1}{3000} + \frac{0.01}{15} + \frac{10^{-3}}{0.37} \ \text{W}^{-1} \cdot \text{m}^2 \cdot \text{K}$$

$$k_c = 270 \ \text{W} \cdot \text{m}^{-2} \cdot K^{-1}$$

$$F = \frac{\pi \cdot 1.5 \cdot 1.8}{2} = 4.2 \text{ m}^2$$

$$\bar{T} = T_c + \frac{T_o - T_c}{\exp \dfrac{k_c F}{\phi_p c_s}} = 32 + \frac{80.4 - 32}{\exp \dfrac{270 \cdot 4.2 \cdot 3600}{1000 \cdot 1440}} = 35°\text{C}$$

Heat balance ($J \cdot hr^{-1}$)

Melt cooling	1000 · 1710(90–80.4)	= 16,416,000
Melt crystallization	1000 · 148,100	= 148,100,000
Flake cooling	1000 · 1440(80.4–35)	= 65,376,000 +
		229,892,000

That is, 64 kW.

The temperature of 28 $m^3 \cdot hr^{-1}$ of cooling water increases from 31 °C to 33°C (28,000 · 2 · 4.2 = 235,200 $kJ \cdot hr^{-1}$).

Review

Drum diameter 1.5 m
Drum length 1.80 m
Immersion depth 300 mm
Rotational speed 1.92 min^{-1}
Production 1000 $kg \cdot hr^{-1}$
Heat load 64 kW
Cooling water flow 28 $m^3 \cdot hr^{-1}$
Flake thickness 1 mm
Flake temperature 35°C

Notes

1. For the solidification of 0.95 mm, 9.2 sec is calculated. An iterative approach is required as an assumption for α_o must be made. The assumption needs verification. Furthermore, the effect of the entrainment has to be checked. The value 9.55 sec is calculated by using the design method for belt coolers. The latter method is straightforward and gives a similar result.
2. The assumption of α_c-values greater than 3000 $W \cdot m^{-2} \cdot K^{-1}$ hardly changes the results. The reason is that $3000 \gg 0.37/0.001$ (λ_s/δ_{act}).

4.11. SCALING UP OF DRUM FLAKER RESULTS

It is required to design the same drum flaker for naphthalene as treated in Sec. 4.10. The results of one of the nine experiments discussed in Sec. 4.9.2 should be scaled up.

Trial No. 8 in Tables 4.2 and 4.3 is selected.

First, the material balance of this experiment is checked.

$$w = \frac{\pi \cdot 0.5 \cdot 2}{60} = 0.052 \text{ m} \cdot \text{sec}^{-1}$$

$$3600 \cdot 0.052 \cdot 1150 \cdot 0.5 \cdot 10^{-3} = 107.6 \text{ kg} \cdot \text{hr}^{-1}$$

Actually collected in 2 min: 3.6 kg (108 $\text{kg} \cdot \text{hr}^{-1}$).

The contact time is 8.8 sec.

The immersion depth of the large flaker is $3 \cdot 100 = 300$ mm.

Hence the "wetted length" is 1.39 m.

One full rotation lasts $8.8 \cdot \pi \cdot 1.5/1.39 = 30$ sec.

The rotational speed is 2.0 min^{-1}, $w = 0.157 \text{ m} \cdot \text{sec}^{-1}$.

$$L = \frac{1000}{3600 \cdot 0.9 \cdot 0.157 \cdot 1150 \cdot 10^{-3}} = 1.71 \text{ m, say } 1.75 \text{ m}$$

The expected flake discharge temperature is 39°C. This design is close to the design arrived at in Sec. 4.10. The latter design required the full set of physical data of naphthalene and α_k-data.

The cooling water flow needed can be calculated when the heat of crystallization of naphthalene is known and assumptions regarding the specific heat are made (e.g., 2000 $\text{J} \cdot \text{kg}^{-1} \cdot \text{K}^{-1}$). One implicit assumption in this procedure is that the heat transfer coefficient α_k is equal on both scales. This is not too bold an assumption. Regarding this issue, the straightforward scale-up procedure is not inferior to the scale-up model as reliable α_k-data are scarce.

4.12. PREGER'S APPROACH

Preger (1968) developed an approach to predict the performance of existing drum flakers on changing the conditions. This method does not require all physical data of the material processed. First, a single-variable study of the rotational speed is carried out. This study provides an adequate basis for predicting the performance on varying:

- The rotational speed
- The cooling water temperature
- The "wetted length".

It was found that the layer thickness δ and hence also the production ϕ_p can be expressed as power law functions of the rotational speed (all independent process variables but the rotational speed must be kept constant). This was discussed in Chapter 2.

$$\delta = An^{(z-1)} \text{ mm} \tag{2.4}$$

z can vary from 0.4 to 0.6. It follows that:

$$\delta = Bt^{(1-z)} \text{ mm} \tag{4.5}$$

It also follows that:

$$\phi_p = Cn^z \text{ kg} \cdot \text{hr}^{-1} \tag{4.6}$$

The temperature of the flakes can also be expressed as a power law function of the rotational speed:

$$\overline{T} = Dn^y \,^\circ\text{C} \tag{4.7}$$

Furthermore, it was found that the next relation is applicable:

$$\phi_p = E\left(\frac{T_o - T_{c1}}{c}\right)^{(1-z)} \text{ kg} \cdot \text{hr}^{-1} \tag{4.8}$$

The latter equation describes the effect of changing the temperature of the entering cooling water (all the other independent process variables must be kept constant).

c is defined by:

$$c(T_f - \overline{T}) = c_l(T_f - T_o) + i + c_s(T_o - \overline{T}) \text{ J} \cdot \text{kg}^{-1}$$

Next, the drum flaker can be considered as an indirect heat exchanger. A series of equations is given.

$$Q = \phi_w c_w(T_{c1} - T_{c2}) \text{ J} \cdot \text{hr}^{-1} \tag{4.9}$$

$$Q = \phi_p c(T_f - \overline{T}) \text{ J} \cdot \text{hr}^{-1} \tag{4.10}$$

$$Q = kF\Delta T_m \text{ W} \tag{4.11}$$

It is assumed that the process fluid and the cooling water flow in parallel. It helps to define an auxiliary variable σ:

$$\sigma = \frac{1 - e^{-\left(1 + \frac{\phi_p c}{\phi_w c_w}\right)\frac{kF}{\phi_p c}}}{1 + \frac{\phi_p c}{\phi_w c_w}} \tag{4.12}$$

It can be shown that the following equations hold:

$$\overline{T} = T_f - \sigma(T_f - T_{c1}) \,^\circ\text{C} \tag{4.13}$$

$$T_{c2} = T_{c1} + \sigma(T_f - T_{c1})\frac{\phi_p c}{\phi_w c_w} \,^\circ\text{C} \tag{4.14}$$

TABLE 4.6 Drum Flaker Measurements

n (min^{-1})	ϕ_p ($kg \cdot hr^{-1}$)	δ_{act} (mm)	T_f (°C)	T_{act} (°C)	ϕ_w ($kg \cdot hr^{-1}$)	T_{c1} (°C)	T_{c2} (°C)
3	310	0.63	130	53	1800	15	23
6	450	0.46	130	65	1800	15	25.5
12	650	0.33	130	80	1800	15	28

Source: Preger, 1968.

Preger (1968) discusses an example of a melt solidifying at 115°C. The area of the drum flaker is 1 m^2. The results of the single-variable study of the rotational speed are given in Table 4.6.

From Preger's results, it is possible to obtain:

$$\delta = 1.052 \cdot n^{-0.465} \text{ mm}$$

$$\phi_p = 172.3 \cdot n^{0.535} \text{ kg} \cdot \text{hr}^{-1}$$

$$\overline{T} = 38.25 \cdot n^{0.297} \text{ °C}$$

Further results of calculations are summarized in Table 4.7. The heat loads were calculated from the cooling water consumption.

Changing the conditions of the 1-m^2 drum flaker is attempted now. First, it is required to calculate the production, the layer thickness, and the flake temperature for $n = 4$ min^{-1} and $n = 15$ min^{-1}. All the other independent process variables are kept constant. The results of the calculations are summarized in Table 4.8.

Second, it is required to calculate the production, the layer thickness, the flake temperature, and the cooling water outlet temperature on adjusting the rotational speed at 8 min^{-1} and the cooling water inlet temperature at 28°C. Again, the other independent process variables are kept constant.

A rotational speed of 8 min^{-1} combined with $T_{c1} = 15$ °C would lead to: $\phi_p = 525$ $kg \cdot hr^{-1}$, $\delta = 0.40$ mm, and $\overline{T} = 70.9$°C.

TABLE 4.7 Results of Drum Flaker Calculations Directly Based on Drum Flaker Measurements

n (min^{-1})	δ_{act} (mm)	T_{act} (°C)	Q ($kJ \cdot hr^{-1}$)	c ($kJ \cdot kg^{-1} \cdot K^{-1}$)	ΔT_m (K)	kF ($kJ \cdot hr^{-1} \cdot K^{-1}$)
3	0.63	53	60,336	2.53	63.3	953.2
6	0.46	65	79,191	2.71	70.7	1120.1
12	0.33	80	98,046	3.02	79.4	1234.8

TABLE 4.8 Results of Drum Flaker Calculations Using Power Law Equations

n (min^{-1})	ϕ_p ($kg \cdot hr^{-1}$)	δ_{pred} (mm)	$\overline{T}$ (°C)
4	360	0.55	58
15	735	0.30	85

Linear interpolation leads to $c = 2.81\ kJ \cdot kg^{-1} \cdot K^{-1}$ (Table 4.7).

When the cooling water inlet temperature is raised to 28°C, the flake temperature increases. c increases also and a value of $2.97\ kJ \cdot kg^{-1} \cdot K^{-1}$ is assumed. The application of Eq. (4.8) leads to:

$$\phi_p = 525\left(\frac{115-28}{115-15} \cdot \frac{2.81}{2.97}\right)^{0.465} = 480\ kg \cdot hr^{-1}$$

$$\delta = 0.63 \cdot \frac{3}{8} \cdot \frac{480}{310} = 0.37\ mm\ \text{(see Table 4.6)}$$

Linear interpolation (based on δ_{act}) results in $kF = 1200\ kJ \cdot hr^{-1} \cdot K^{-1}$ (Table 4.7).

On applying Eq. (4.12), $\sigma = 0.5320$ can be calculated.

Furthermore, it follows from Eqs. (4.13) and (4.14) that $\overline{T} = 75.7$°C and $T_{c2} = 38.2$°C. Now it can be checked whether the assumption $c = 2.97\ kJ \cdot kg^{-1} \cdot K^{-1}$ was correct.

$$c = \frac{1800 \cdot 4.2(38.2-28)}{480(130-75.7)} = 2.96\ kJ \cdot kg^{-1} \cdot K^{-1}$$

The correctness of the interpolation to arrive at $kF = 1200\ kJ \cdot hr^{-1} \cdot K^{-1}$ can also be checked.

$$\Delta T_m = \frac{(130-28)-(75.7-38.2)}{{}^{e}\log\frac{130-28}{75.7-38.2}} = 64.5\ K$$

$$kF = \frac{1800 \cdot 4.2(38.2-28)}{64.5} = 1196\ kJ \cdot hr^{-1} \cdot K^{-1}$$

Third, it is required to calculate the production and the flake thickness when the "wetted length" increases by 50% ($8\ min^{-1}$, $T_{cl} = 15$ °C).

$$\phi_p = 525 \cdot 1.5^{0.465} = 635\ kg \cdot hr^{-1}$$

$$\delta = 0.40 \cdot 1.5^{0.465} = 0.48\ mm$$

According to Preger's model, increasing the "wetted length" by 50% has the same effect on the thickness of the layer as decreasing the rotational speed by 50%. This makes sense, as the contact time is the same in both cases. Preger's model thus does not take into account differences in entrainment.

Note

Preger's approach is helpful to model the performance of a specific drum flaker on processing a specified material. The cooling water flow and the way the cooling water is applied must also be kept constant. The independent process variables that can be varied are:

- The rotational speed
- The "wetted length"
- The cooling water inlet temperature.

Their impact on the following dependent process variables can be studied:

- The flake thickness
- The production
- The flake temperature.

The study starts with the execution of a single-variable study of the rotational speed of the flaking drum.

4.13. VERIFICATION OF SOME OF PREGER'S GENERALIZATIONS

Sec. 4.9.3 contains measurements by Kloppenburg (1968). The production was also measured during these trials. The measurements of the first series of trials are reviewed in Table 4.9. The data of the first four columns of Table 4.9

TABLE 4.9 Experimental Verification of Two of Preger's Power Law Equations for the Flaking of a Product

Test No.	n (min^{-1})	δ_{act} (mm)	ϕ_P ($kg \cdot hr^{-1}$)	$\delta = 0.80 \cdot n^{-0.53}$ (mm)	$\phi_P = 600 \cdot n^{0.47}$ ($kg \cdot hr^{-1}$)
1	1.34	0.67	725	0.69	688
2	1.83	0.62	800	0.58	797
3	2.40	0.49	880	0.50	905
4	3.26	0.445	1045	0.43	1046
5	4.38	0.360	1170	0.37	1201

See Table 4.4.

can also be found in Table 4.4. The results can be described by the following power law equations:

$$\delta = 0.80 \cdot n^{-0.53} \text{ mm (see Fig. 4.10)}$$
$$\phi_p = 600 \cdot n^{0.47} \text{ kg} \cdot \text{hr}^{-1} \text{ (see Fig. 4.11)}$$

Table 4.9 also contains the results of the calculations. For equal contact times, the flake thicknesses stated in Table 4.4 are slightly larger than the flake thicknesses stated in Table 4.5. This can be explained by the greater circumferential velocities at the trials in Table 4.4 resulting in more entrainment.

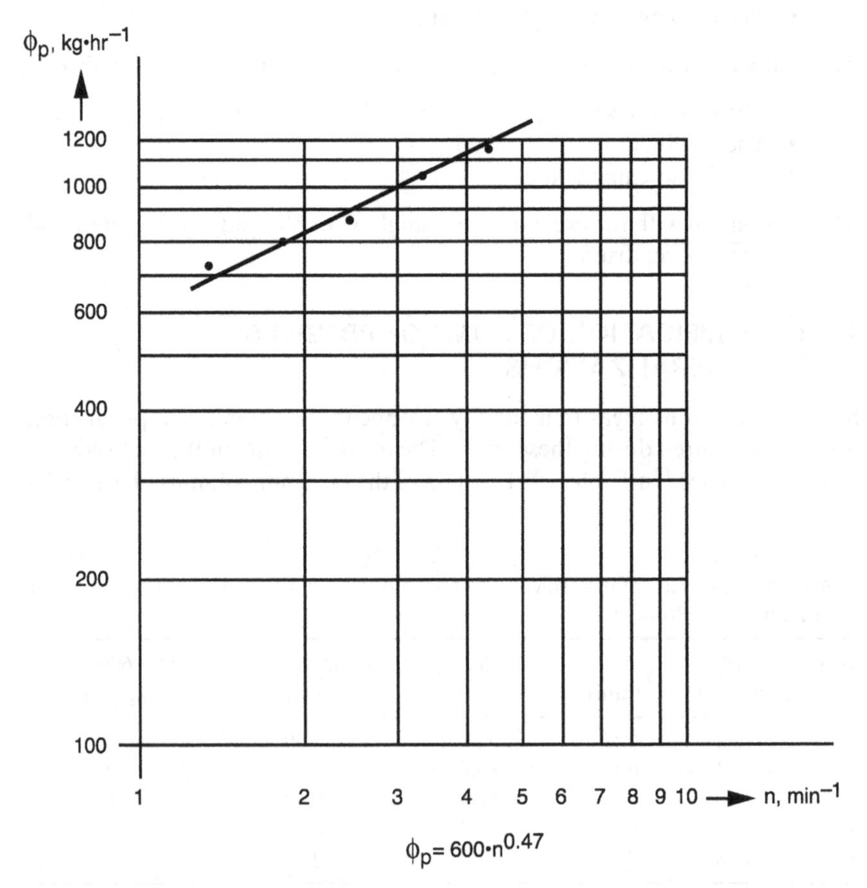

FIG. 4.11 A Preger plot (ϕ_p).

TABLE 4.10 Experimental Verification of Three of Preger's Power Law Equations for Naphthalene Flaking

Trial No.	n (min^{-1})	δ_{act} (mm)	ϕ_p (kg · hr^{-1})	T_{act} (°C)	$\delta=1.55 \cdot n^{-0.57}$ (mm)	$\phi_p=84 \cdot n^{0.43}$ (kg · hr^{-1})	$\overline{T}=26 \cdot n^{0.3}$ (°C)
6	5	0.60	170.3	41.4	0.62	168	42
7	3	0.85	135	35.7	0.83	135	36
8	2	1.0	108	38.5	1.04	113	32
9	3.6	0.8	151.5	41.4	0.75	146	38

See Tables 4.2 and 4.3.

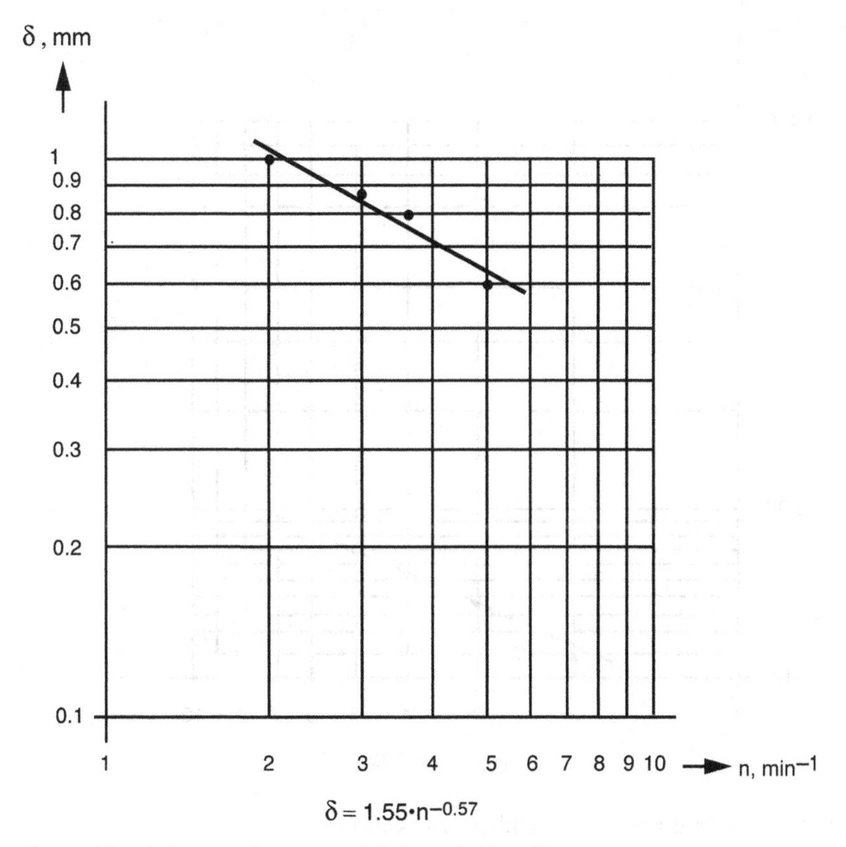

FIG. 4.12 A Preger plot for naphthalene flaking (δ).

According to Preger's model, these flake thicknesses should be identical.

Sec. 4.9.2 contains measurements regarding the flaking of naphthalene. The measurements of trial Nos. 6, 7, 8, and 9 are reviewed in Table 4.10. The data of the first five columns can also be found in Tables 4.2 and 4.3. The data can be described by the following power law equations:

$\delta = 1.55 \cdot n^{-0.57}$ mm (see Fig. 4.12)
$\phi_p = 84 \cdot n^{0.43}$ kg · hr^{-1} (see Fig. 4.13)
$T = 26 \cdot n^{0.3}$ °C (see Fig. 4.14)

Table 4.10 also contains the results of calculations.

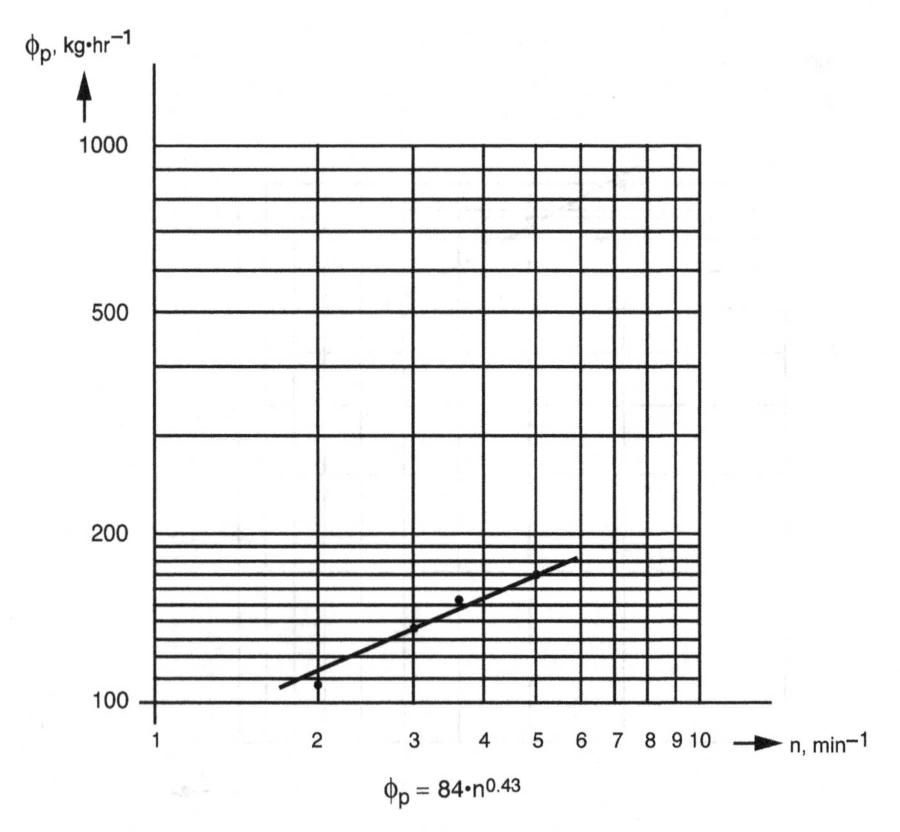

FIG. 4.13 A Preger plot for naphthalene flaking (ϕ_p).

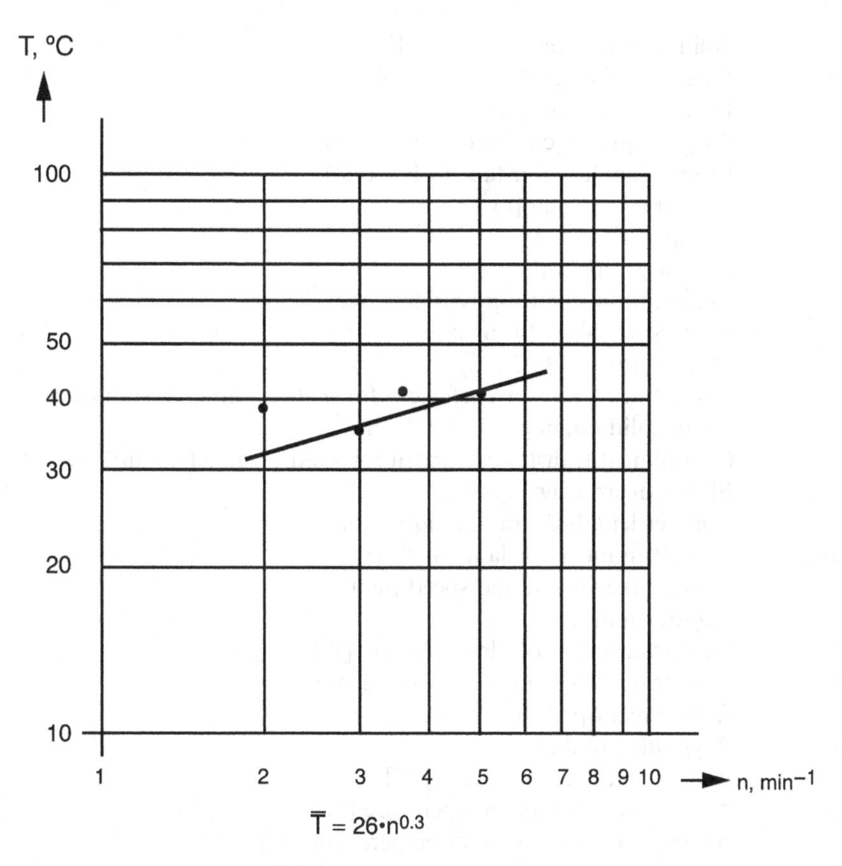

FIG. 4.14 A Preger plot for naphthalene flaking (T).

The flaking of six materials was investigated by Preger (1970, 1980). It appeared that the exponent z in Eq. (4.6) had values in the range 0.4–0.6.

LIST OF SYMBOLS

A	Proportionality constant in Eq. (2.4)
a	Flaking efficiency factor (0.0–1.0)
B	Proportionality constant in Eq. (4.5)
C	Proportionality constant in Eq. (4.6)
c	Apparent specific heat [$J \cdot kg^{-1} \cdot K^{-1}$ or $kJ \cdot kg^{-1} \cdot K^{-1}$]
c_l	Melt specific heat [$J \cdot kg^{-1} \cdot K^{-1}$]

c_s	Solid specific heat $[J \cdot kg^{-1} \cdot K^{-1}]$
c_w	Water specific heat $[J \cdot kg^{-1} \cdot K^{-1}]$
D	Flaker drum diameter [m]
	Proportionality constant in Eq. (4.7)
E	Proportionality constant in Eq. (4.8)
F	Heat transfer area $[m^2]$
F_0	Gregorig function
F_1	Gregorig function
g	Acceleration due to gravity $[m \cdot sec^{-2}]$
h	Immersion depth [m or mm]
i	Heat of fusion $[J \cdot kg^{-1}]$
k	Overall heat transfer coefficient for melt cooling, crystallization, and solid cooling $[kJ \cdot hr^{-1} \cdot m^{-2} \cdot K^{-1}]$
k_c	Overall heat transfer coefficient for solid cooling $[W \cdot m^{-2} \cdot K^{-1}]$
L	Flaker drum length [m]
l	Contact length drum/pan liquid [m]
Nu_{lam}	Nusselt number for laminar flow
n	Drum flaker rotational speed $[min^{-1}]$
Pr	Prandtl number
Q	Heat transferred on drum flaking [W]
Q_c	Heat transferred on solid cooling [W]
q	Entrainment $[m^3 \cdot m^{-1} \cdot sec^{-1}]$
Re	Reynolds number
$\overline{T}$	Average flake temperature [°C]
T_{act}	Actual average flake temperature [°C]
T_c	Average cooling medium temperature [°C]
T_{c1}	Cooling water entrance temperature [°C]
T_{c2}	Cooling water exit temperature [°C]
T_f	Melt temperature [°C]
T_o	Melting point [°C]
ΔT_m	Logarithmic mean temperature difference [K]
t	Crystallization time [sec]
	Time for one rotation [sec]
t_{act}	Actual crystallization time [sec]
t_e	Time for an experiment [min]
w	Drum flaker circumferential velocity $[m \cdot sec^{-1}]$
y	Exponent in Eq. (4.7)
z	Exponent/part of an exponent in Eqs. (2.4), (4.5), (4.6), and (4.8)
α_c	Heat transfer coefficient (medium-side) $[W \cdot m^{-2} \cdot K^{-1}]$
α_k	Heat transfer coefficient (medium-side, includes metal wall resistance) $[W \cdot m^{-2} \cdot K^{-1}]$

α_o	Heat transfer coefficient (process-side) $[W \cdot m^{-2} \cdot K^{-1}]$
δ	Product layer thickness [m or mm]
δ_{act}	Actual product layer thickness [mm]
δ_{ent}	Thickness of the entrained layer [mm]
δ_{pan}	Thickness of the layer grown in the pan [mm]
δ_{pred}	Predicted layer thickness [mm]
δ_w	Metal wall thickness [m or mm]
θ	Dimensionless number
λ_l	Melt thermal conductivity $[W \cdot m^{-1} \cdot K^{-1}]$
λ_s	Solid thermal conductivity $[W \cdot m^{-1} \cdot K^{-1}]$
λ_w	Metal wall thermal conductivity $[W \cdot m^{-1} \cdot K^{-1}]$
μ	Dimensionless number ("sensible heat/latent heat")
	Dynamic viscosity $[N \cdot sec \cdot m^{-2}]$
μ'	Dimensionless number ("sensible heat/latent heat" for "superheated" melt)
ν	Melt kinematic viscosity $[m^2 \cdot sec^{-1}]$
ρ_l	Melt specific mass $[kg \cdot m^{-3}]$
ρ_s	Solid specific mass $[kg \cdot m^{-3}]$
σ	Auxiliary variable {see Eqs. (4.12), (4.13), and (4.14)}
τ	Dimensionless number ("time")
τ'	Dimensionless number ("time" for "superheated" melt)
ϕ_p	Flaker production $[kg \cdot hr^{-1}]$
ϕ_w	Cooling water flow $[kg \cdot hr^{-1}]$
χ	Dimensionless number ("layer thickness")

REFERENCES

Groot Wassink, J. (1976). *Process analysis and design of drum flakers*. The Hague, The Netherlands: Stichting Nederlandse Apparaten voor de Procesindustrie.

Groot Wassink, J. (1977). Description and design of drum flakers. *Aufbereitungs-Technik* 18:85. In German.

Kloppenburg, W. (1968). Private communication.

Preger, M. (1968). Drum flaker performance on changing the operating conditions. *Aufbereitungs-Technik* 9:123. In German.

Preger, M. (1969). Drum flakers as heat exchangers. *Maschinenmarkt* 75:1622. In German.

Preger, M. (1970). Drum flaker applications and design. *Aufbereitungs-Technik* 11:551. In German.

Preger, M. (1980). Drum flaker systems: status of the development and trends. *Maschinenmarkt* 86:80. In German.

VDI (1993). *VDI Heat Atlas*. Düsseldorf, Germany: VDI Verlag.

APPENDIX A DATA FOR 29 CAPROLACTAM DRUM FLAKING TRIALS

Experiment No.	1	2	4	5	6	7	8	9	10	11	12	14	15	16
Independent variables														
n (min^{-1})	7.5	6	7.0	3.0	3.0	4.0	4.0	3.5	2.6	1.0	1.0	7.0	10.0	2
h (mm)	20	100	100	100	100	100	100	100	100	100	25	10	10	100
T_f (°C)	95	95	86.5	84	85	91	91	80	80	80	80	80	80	86
T_c (°C)	33	34	24	21	23.5	23.5	27.5	18	18.5	18	17	23	22	23
t_e (min)	4	9	2	6	5	3	3	3	3	4	15	2	2.5	1
Dependent variables														
δ_{act} (mm)	0.430	0.484	0.48	0.65	0.58	0.55	0.55	0.62	0.65	1.1	0.90	0.42	0.45	0.60
T_{act} (°C)	48	45	39	29	28	35	36	26	24	22	20	37.8	67.5	30
ϕ_p (kg · hr^{-1})	166.5	154	180	105	93.6	120	120	120	96	54	44.4	168	266.4	72
l (m)	0.2	0.46	0.46	0.46	0.46	0.46	0.46	0.46	0.46	0.46	0.23	0.14	0.14	0.46
t_{act} (sec)	1.0	2.9	2.5	5.9	5.9	4.4	4.4	5.0	6.8	17.6	8.8	0.76	0.53	8.8

Experiment No.	17	18	19	20	21	22	23	24	25	26	27	28	29	30	31
Independent variables															
n (min^{-1})	3	5	3	2	4	4	2	2	3	3	3	3	8.5	9	8.5
h (mm)	100	100	100	30	30	100	100	140	140	140	140	140	100	100	100
T_f (°C)	80	80	80	80	80	80	80	80	80	80	80	80	78	77.4	76.5
T_c (°C)	7.9	2.25	12	12	11.75	13.25	13.65	13.8	0	6.5	16.2	20.75	7	7.5	7.5
t_e (min)	4	3	5	3	3	3	3	3	3	3	3	4	2	2	4
Dependent variables															
δ_{act} (mm)	0.73	0.64	0.72	0.62	0.49	0.646	0.82	0.92	0.90	0.85	0.81	0.695	0.65	0.63	0.62
T_{act}(°C)	9	19.0	26	19	22.2	24.8	21.3	21.6	19.0	20.8	25	30	34.6	43.2	28.2
ϕ_p (kg · hr^{-1})	127.5	154	116.4	70	108	132	88	98	140	138	134	112.5	260.1	291.3	261
l (m)	0.46	0.46	0.46	0.25	0.25	0.46	0.46	0.56	0.56	0.56	0.56	0.56	0.46	0.46	0.46
t_{act} (sec)	5.9	3.5	5.9	4.8	2.4	4.4	8.8	10.7	7.1	7.1	7.1	7.1	2.1	2.0	2.1

Note: Experiments Nos. 3, 13, and two further experiments were excluded.

5

Prilling

5.1. INTRODUCTION

Prilling as a process is defined and discussed in Sec. 5.2. The generation of droplets is treated in Sec. 5.3. The next four sections contain the elements for a design method. First, the terminal velocity of a falling sphere is dealt with in Sec. 5.4. Second, the heat transfer from a falling sphere is the subject of Sec. 5.5. The solidification of a sphere is considered in Sec. 5.6. The solidification process occurs as a result of instationary heat transfer from the sphere to the air. The next section covers the cooling of the crystallized (solidified) material. The design method is reviewed in Sec. 5.8. Sec. 5.9 contains the verification of the recommended design method. A design example is discussed in the last section.

5.2. GENERAL DESCRIPTION

Prilling is the process of spray crystallization. A liquid is sprayed to produce drops falling through a cooling medium and crystallizing into particles.

Surface tension causes the liquid drops to adopt a spherical shape as this results in the smallest surface-area-to-volume ratio. Prilling was used originally for the production of lead shot (lead has a melting point of 327.4°C). In the recent past, the process was used in the fertilizer industry for the production of urea and ammonium nitrate. Today, prilling the latter two compounds has largely been replaced by fluid bed granulation. The main reason is that the latter technique produces stronger and more abrasion-resistant granules. Concerning urea and ammonium nitrate, prilling and fluid bed granulation are compared in Secs. 8.3 and 8.4. Inorganic substances such as sodium hydroxide (see Figs. 5.1 and 5.2) and iodine and certain organic compounds (like stearic acid) are still prilled. Prilling as a form of granulation (sometimes called congealing) is cost-effective for large outputs. In principle, the particles have a uniform size. There is an upper limit to prill size because of practical limitations to prill tower height, as, for example, the terminal velocity of a 500-μm ammonium nitrate prill is 3 $m \cdot sec^{-1}$. The process of crystallization and cooling takes a number of seconds. Congealing towers can have a maximum free-fall height of 60 m (overall height of about 100 m), while the tower diameter may be about 15 m. A further aspect of prill size is that large droplets do not assume or cannot retain a spherical form; for example, raindrops become flat when $Re > 10^3$ (Grassmann and Reinhart, 1961). $Re = 10^3$ corresponds with $d_p \approx 3$ mm. Urea has a viscosity of $2.6 \cdot 10^{-3}$ $N \cdot sec \cdot m^{-2}$ at 150°C and is thus hardly more viscous than water.

Moderately viscous melts can be congealed as well as low-viscosity melts. For instance, the prilling of a melt having a viscosity of 0.8 $N \cdot sec \cdot m^{-2}$

FIG. 5.1 NaOH prills 0.5–1.2 mm.

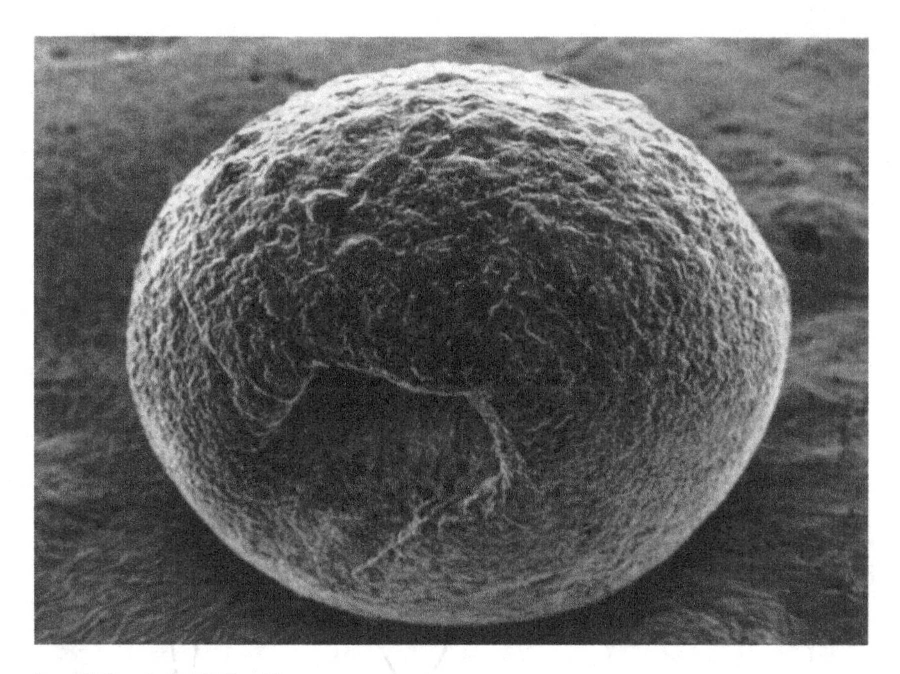

Fig. 5.2 A NaOH prill.

is practiced industrially. Beyond being compared to fluid bed granules, prills can also be compared to granules made by means of drum granulation or pan granulation. These granules are also stronger and more abrasion-resistant than prills.

For example, granulated urea has a 3–6 lb ASTM crushing strength while prilled urea has a strength of approximately 2 lb. Particles passing through an 8-mesh (2380-μm) screen and retained by a 10-mesh (1680-μm) screen were compared (Ruskan, 1976). Prills are generally too small for use in blending with other granular materials.

Fig. 5.3 exhibits a typical prilling plant. The melt is stirred in a vessel at a temperature 10–20 K higher than the melting point. Lower temperatures increase the risks of lines choking due to crystallization, while higher temperatures result in increased cooling duty. To avoid discoloration, the melt can be blanketed with nitrogen.

A pump transfers the liquid to the top of the tower. Here the melt is sprayed and the droplets fall down. The device that generates the spray is the heart of the prilling operation. The generation of droplets will be discussed in Sec. 5.3.

Prilling towers are usually constructed in concrete (fertilizers), steel, or stainless steel.

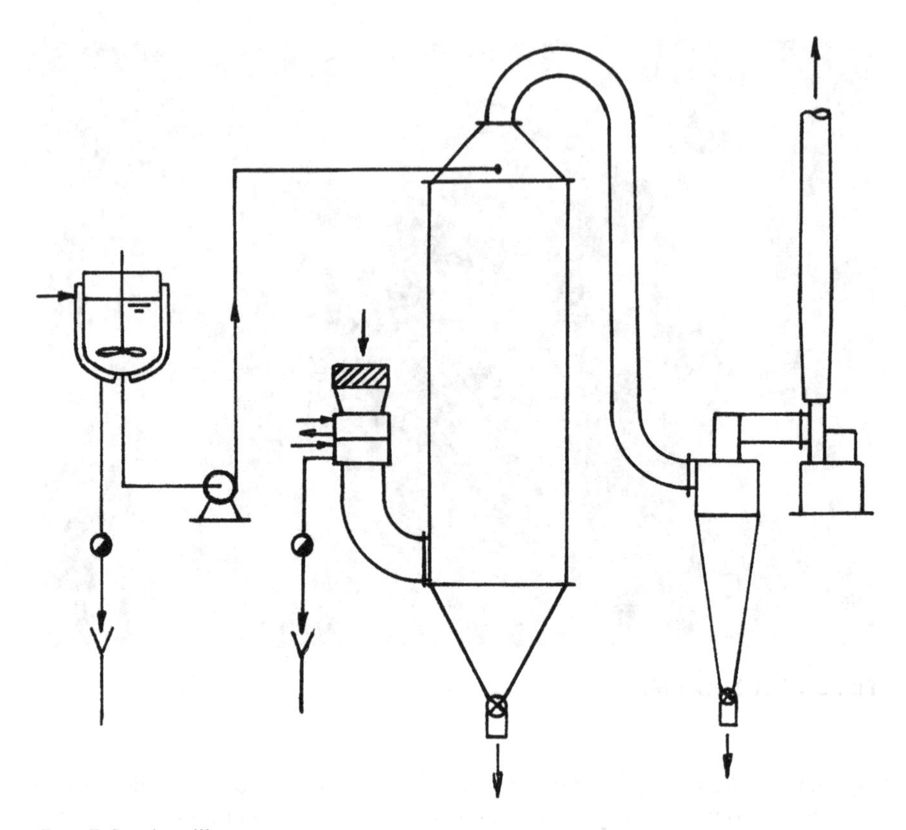

FIG. 5.3 A prill tower.

The superficial gas velocity of the cooling medium (usually air) is in the range 0.5–2 $m \cdot sec^{-1}$, and the temperature increases by 20 K in many instances. These two rules-of-thumb fix the congealing tower's diameter. Furthermore, there should be little impingement on the tower walls as this can eventually result in the breaking off of chunks which will plug-up the tower bottom. Spray crystallization can occur in a cocurrent mode or in countercurrent operation. Cocurrent operation results in quick solidification because the melt is contacted with cold air. Countercurrent operation tends to raise the residence time of the particles.

Normally, the prills are collected at the base of the tower in a conical hopper and removed by a vibratory conveyor, a belt conveyor, or a pneumatic transport system. A tower having a flat base is a different option. Now a rotating rake feeds the product onto a conveyor below the tower. A flat base saves height (hoppers can have heights of, e.g., 20 m) but can cause breakage.

Often, some drying (water evaporation) can be accomplished while the material crystallizes and cools down. However, because of the short residence time (seconds), complete drying is often not possible. For instance, there are ammonium nitrate plants in which a concentrated solution containing 5% water by weight is pumped to the top of a prilling tower. The prills leaving the tower still contain 4% water by weight. Postdrying is necessary to prevent the occurrence of polymorphic transitions. Furthermore, often, the prills have to be cooled in, e.g., contact coolers.

Usually, for environmental reasons, there is underpressure in the tower. Often, inlet filters are installed to retain suspended solid particles. Such filters separate, e.g., 99.9% by weight of all particles larger than 2 μm. In Fig. 5.3, the air can be either cooled or heated. Cooling can be required by the process, while heating is useful to dry the tower when cleaning has taken place. In some cases, it is necessary to use conditioned air (control of both temperature and relative humidity).

The mechanism by which the droplets are formed will be discussed in detail in Sec. 5.3. Generally, the drop size is uniform in practice. However, often, the larger drops are interspersed by smaller satellite drops. These small drops can give rise to an effluent problem. The exhaust air passes through a cyclone in Fig. 5.3.

Particles having a diameter of approximately 10 μm are retained with an efficiency of 50%. Often, it is necessary to filter the air before venting it to the atmosphere. The quality of the exhaust air must comply with the relevant regulations. In many instances, the prill towers must have wet scrubbing equipment. The treatment of the exhaust air can have important consequences because the airflows are large.

Sometimes milled solid material is added to the tower. This fine material acts as seeds.

Small-scale tests can only be carried out in high equipment as the particles travel under the action of gravity. It is recommended to check the possibility of a dust explosion occurring in a tower as dust explosions have occurred in prilling towers in the past. The "miniprills" referred to earlier may give problems. Filters should receive special attention as they collect the fine material.

5.3. FORMATION OF DROPLETS

5.3.1. Introduction

Generally, a liquid-in-gas dispersion can be generated by three different methods:

- A pressure nozzle
- A rotating perforated bucket or cup
- A two-fluid nozzle.

Pressure nozzles and rotating devices are widely used for prilling. Prills having diameters in the range 0.5–1.5 mm are generated. Both pressure nozzles and rotating devices can be used for liquids having viscosities up to $1\ N \cdot sec \cdot m^{-2}$. Perforated buckets are depicted in Fig. 5.4. A rotating cup generating caustic soda droplets is shown in Fig. 5.5. Two-fluid nozzles (air atomization) are used for relatively small production capacities. In addition, the droplets are relatively fine (<0.5 mm).

5.3.2. Pressure Nozzles

At low flow rates, drops form at the nozzle tip. On raising the flow rate, a jet is obtained. At a certain distance from the nozzle, the jet breaks up into droplets. The length of a jet emerging from a given nozzle is a function of the flow rate only if the liquid properties are kept constant. The jet length increases on increasing the velocity up to a maximum at the critical velocity. The jet length decreases when the flow rate is raised further until the jet breakup point reaches the nozzle.

FIG. 5.4 Perforated cones for the distribution of the melt. (Courtesy of Kreber, Vlaardingen, The Netherlands.)

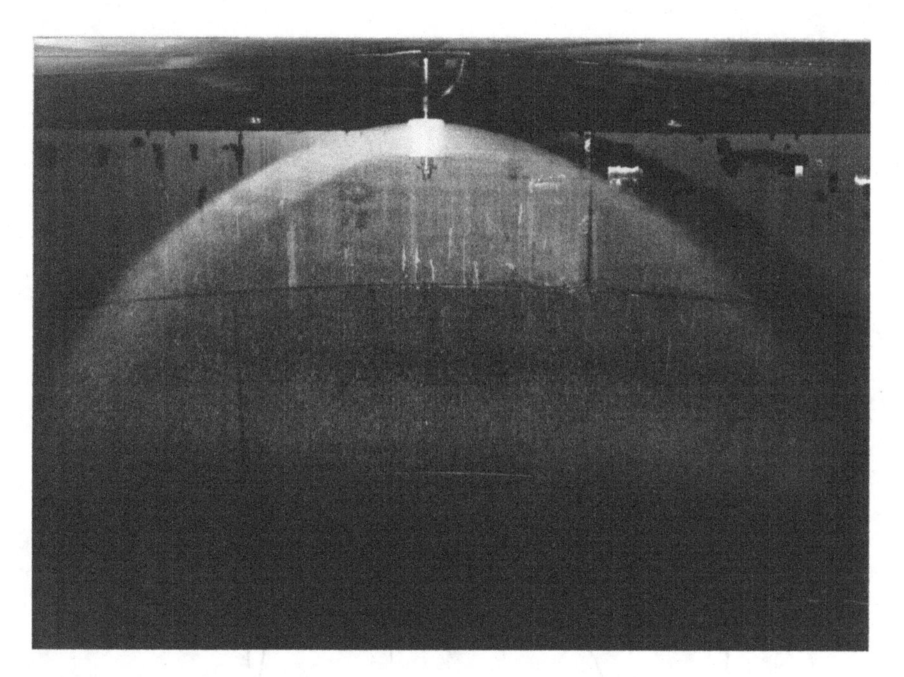

FIG. 5.5 A rotating cup generating caustic soda droplets.

Fig. 5.6 is a qualitative plot of jet length vs. flow rate. Rayleigh (1878) made an analytical study of the collapse of a liquid jet. Rayleigh stated that jet instability could be attributed to surface tension forces. It was postulated that an unstable equilibrium would be destroyed by disturbances. In principle, these disturbances might have all frequencies and hence all wavelengths. In nonviscous liquids, the effect of a disturbance characterized by $\lambda/d_h = 4.508$ would increase quickest and break up the jet. When a cylinder having a length of $4.508 \cdot d_h$ and a diameter d_h contracts into a spherical droplet, the droplet's diameter will be:

$$d_p = 1.89 \cdot d_h \tag{5.1}$$

Generally, the breakup of jets having a velocity smaller than the critical velocity is called Rayleigh breakup. Von Ohnesorge (1936), Duffie and Marshall (1953), and Marshall (1954) photographed such disintegrating jets. It can be seen that axially symmetric disturbances produce breakup. The drops are interspersed by small satellite drops.

Merrington and Richardson (1947), Duffie and Marshall, and Van den Berg and Hallie (1960) documented that the average droplet size due to

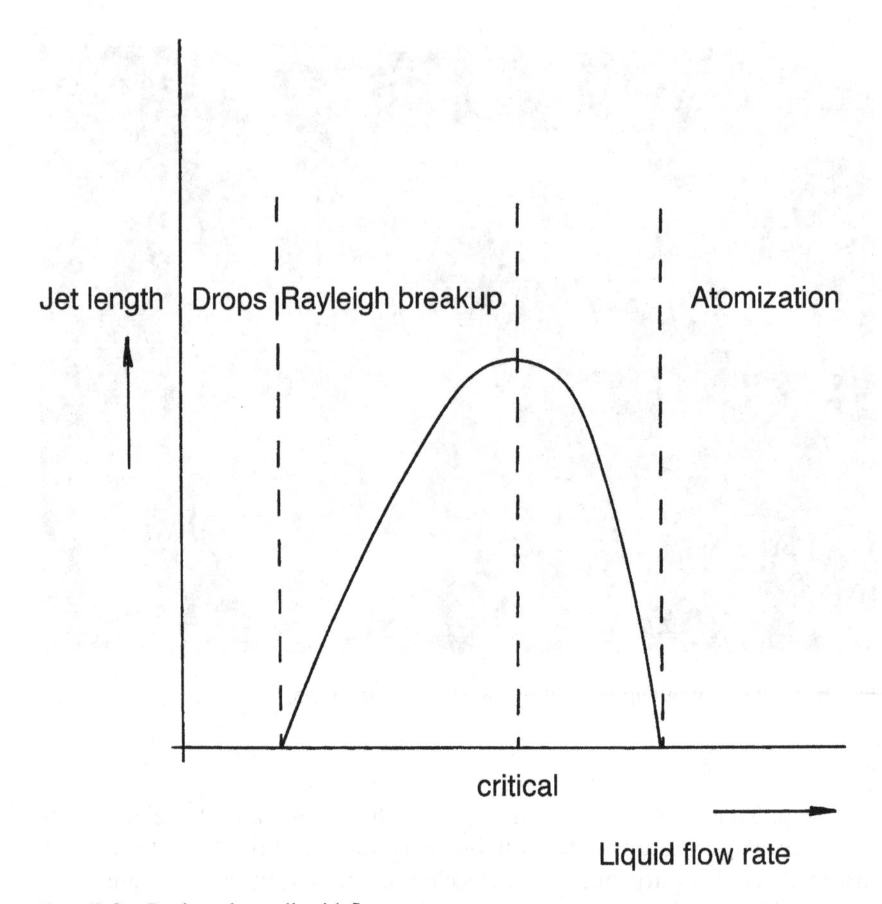

FIG. 5.6 Jet length vs. liquid flow rate.

Rayleigh breakup is approximately twice the nozzle size. This will be discussed further in this section.

The breakup of jets having a velocity greater than the critical velocity has also been photographed by Von Ohnesorge. Breakup is caused by disturbances which are symmetrical about a helical axis starting at the orifice. Merrington and Richardson and Van den Berg and Hallie found that the average droplet size is, also in this case, approximately twice the nozzle size. This will be discussed in this section as well.

Weber (1931) analyzed the disintegration of a jet mathematically. His analysis extended that of Rayleigh's to include the effect of viscous forces on jet breakup. Weber arrived at $\lambda/d_{\mathrm{h}} = \pi\sqrt{2}\left(1 + 3\mu_{\mathrm{l}}/\sqrt{\rho_{\mathrm{l}}\sigma d_{\mathrm{h}}}\right)^{1/2}$. In this equation, the value of λ/d_{h} reduces to 4.44 when $\mu_{\mathrm{l}} = 0$. This is about the value predicted by Rayleigh for this case.

The diameter of the spherical droplet becomes:

$$d_p = 1.88\left(1 + \frac{3\mu_l}{\sqrt{\rho_l \sigma d_h}}\right)^{1/6} d_h \tag{5.2}$$

Basically, Eq. (5.2) indicates the effect of viscosity on the diameter of the droplets. Many liquid specific masses are in the range 750–1250 $kg \cdot m^{-3}$, while liquid surface tensions are usually in the range 0.02–0.07 $N \cdot m^{-1}$. (However, the surface tensions of liquid metals are much higher, e.g., 0.5 $N \cdot m^{-1}$.) Liquid viscosity can vary between 10^{-4} and 1 $N \cdot sec \cdot m^{-2}$.

Example

Typical liquid data are $\rho_l = 1000\ kg \cdot m^{-3}$ and $\sigma = 0.03\ N \cdot m^{-1}$. Furthermore, $d_h = 10^{-3}$ m.

For $\mu_l = 10^{-3}\ N \cdot sec \cdot m^{-2}$, Eq. (5.2) predicts $d_p = 1.88 \cdot d_h = 1.88$ mm. For $\mu_l = 1\ N \cdot sec \cdot m^{-2}$, the prediction is $d_p = 3.05 \cdot d_h = 3.05$ mm.

Besides taking photographs, Von Ohnesorge presented data concerning the breakup of jets. Merrington and Richardson developed a criterion from these data for atomization at the orifice:

$$\frac{\mu_l}{\sqrt{\rho_l \sigma d_h}} \geq 2000\left(\frac{\mu_l}{\rho_l v_h d_h}\right)^{4/3} \tag{5.3}$$

The left-hand side is constant for a given liquid and nozzle as it contains physical properties and the nozzle size only. The velocity is equal to the nozzle-atomization velocity when the right-hand side is equal to the left-hand side. Von Ohnesorge used nozzles having diameters in the range 0.5–4.0 mm. His liquids had various viscosities. Again, it is possible to get a basic idea of the practical meaning of Eq. (5.3).

Example

Typical liquid data are $\rho_l = 1000\ kg \cdot m^{-3}$ and $\sigma = 0.03\ N \cdot m^{-1}$. Furthermore, $d_h = 10^{-3}$ m. Calculate v_h by using Eq. (5.3). For $\mu_l = 10^{-3}\ N \cdot sec \cdot m^{-2}$, $v_h = 14.5\ m \cdot sec^{-1}$. For $\mu_l = 1\ N \cdot sec \cdot m^{-2}$, $v_h = 80\ m \cdot sec^{-1}$.

Merrington and Richardson found that the prediction of Eq. (5.3) is approximately correct except for a liquid having a high viscosity, i.e., 1 $N \cdot sec \cdot m^{-2}$ (glycerol).

Attention will now be paid to more data concerning the experimental verification of Eqs. (5.2) and (5.3).

Merrington and Richardson reported results for noncrystallizing liquids; for example, 2-mm carbon tetrachloride and water droplets were generated by a 1-mm nozzle if the liquids left the nozzle at velocities in the range 1–10 $m \cdot sec^{-1}$. The two authors found that 3.5-mm glycerol droplets were generated by a 1-mm nozzle (velocity range 1–5 $m \cdot sec^{-1}$).

The droplets were obtained by means of jets fired by gas pressure. Narrow particle size distributions were obtained.

Example

It is required to compare the droplet diameters found by Merrington and Richardson to the droplet diameters calculated by means of Eq. (5.2). Furthermore, it is required to calculate the velocity at which atomization at the nozzle starts by means of Eq. (5.3).

Table 5.1 contains the physical properties according to the two authors and the results of the calculations. The calculated droplet sizes are approximately equal to the droplet sizes found experimentally.

The atomization velocity calculated for glycerol is too high. For very viscous liquids, the Von Ohnesorge criterion appears to indicate too high a velocity for the changeover.

Merrington and Richardson also reported results for nozzle velocities in the range 25–100 $m \cdot sec^{-1} \cdot d_p$ appeared to be approximately inversely proportional to v_h and directly proportional to $v_l^{0.2} \cdot d_p$ was independent of the nozzle diameter and the surface tension. Typical results are reproduced in Table 5.2.

Duffie and Marshall (1953) studied the breakup of liquid jets in air by means of photography. Their jet atomizer consisted of a small-bore capillary tube through which the feed liquid was pumped downward. The jet viscosity was varied between $0.836 \cdot 10^{-3}$ and $36.0 \cdot 10^{-3}$ $N \cdot sec \cdot m^{-2}$ by the use of

TABLE 5.1 Mean Droplet Sizes Produced by Pressure Atomization According to Theory and According to Merrington and Richardson (1947)

	Specific mass ($kg \cdot m^{-3}$)	Dynamic viscosity ($N \cdot sec \cdot m^{-2}$)	Surface tension ($N \cdot m^{-1}$)
Carbon tetrachloride	1600	$0.6 \cdot 10^{-3}$	0.025
Water	1000	$1.2 \cdot 10^{-3}$	0.073
Glycerol	1260	1.0	0.064
	d_p calculated (mm)	d_p experiment (mm)	v_h Eq. (5.3) ($m \cdot sec^{-1}$)
Carbon tetrachloride	1.88	2	8.8
Water	1.88	2	21
Glycerol	2.83	3.5	92

TABLE 5.2 Mean Droplet Sizes Produced by Pressure Atomization According to Merrington and Richardson (1947)

Liquid	v_h (m · sec^{-1})	d_p (mm)
Soap solution	25	1.10
	36	0.55
	57	0.38
	68	0.30
Glycerol	88	0.88
	109	0.73

sucrose solutions; the specific masses varied between 996 and 1269 kg · m^{-3}. The nozzle diameters used were 300, 215, 175, and 145 μm and the jet velocity varied between 1.4 and 8.7 m · sec^{-1}. The results of 19 experimental runs were reported and it appeared that, on average, the geometric mean drop diameter was 2.0 times the nozzle diameter. However, the spread was considerable as the factor varied between 1.4 and 3.1. The factor tended to exceed 2.0 for the more viscous solutions and to become smaller than 2.0 for velocities in the range 5–10 m · sec^{-1}.

Motion pictures of jet breakup were taken at a film speed of 7000 frames per second. Measurements from these pictures indicated an average ratio of disturbance wavelength to jet diameter of approximately 4.6 based on drop spacing. This compares well with Rayleigh's theoretical value of 4.508.

The authors stated that in each experiment, a range of drop sizes was obtained because vibrations of varying and discontinuous ranges of wavelengths, varying in amplitudes and acting in various directions, were impressed on the jets. Thus Rayleigh's theory only held approximately. Van den Berg and Hallie (1960) reported measurements with a highly concentrated urea solution ($\mu_l = 2.15 \cdot 10^{-3}$ N · sec ·m^{-2}). The solution was ejected vertically downward by means of air pressure through a tower having a height of approximately 40 m. The temperature was 140°C and the diameter of the nozzle was 0.5 mm. The authors found that the mean prill diameter was 1.1 mm for velocities in the range 1.25–5 m · sec^{-1}. In contrast to the findings of Merrington and Richardson, Van den Berg and Hallie found that the particle diameter increased when the velocity was raised above 5 m · sec^{-1}; for example, 1.6 mm was measured at 20 m · sec^{-1}. This was attributed to the instantaneous crystallization of the outer shell of the primary drops. Thus further disruption was prevented.

It is concluded that the diameter of the droplet from a pressure nozzle is approximately twice the nozzle diameter if the viscosity is in the range 10^{-3}–10^{-1} N · sec · m^{-2} and the velocity is in the range 1–10 m · sec^{-1}.

The capacity of a pressure nozzle is limited when the liquid flows through at a rate of, e.g., 5 m · sec^{-1}.

Example

Nozzle diameter 1.0 mm
$v_h = 5\ \text{m} \cdot \text{sec}^{-1}$
$\rho_l = 1000\ \text{kg} \cdot \text{m}^{-3}$
Now the hourly production can be calculated.

$$\phi_p = \frac{\pi}{4} \cdot 10^{-6} \cdot 5 \cdot 1000 \cdot 3600 = 14.1\ \text{kg} \cdot \text{hr}^{-1}$$

For 3 t·hr^{-1}, 213 nozzles are required.

5.3.3. Rotating Nozzles

It is possible to obtain large outputs by means of rotating conical or cylindrical perforated cups. The liquid jets are formed under the influence of the centrifugal force. The mechanism of drop formation when using a rotating device is essentially the same as that applying to fixed nozzles. However, the contact with the air in the tower proceeds in a different way.

First, the capacity of a rotating cylindrical cup will be discussed. The melt is fed into the center of the spinning device. Under the influence of the centrifugal force, the liquid will be forced against the perforated wall. For a liquid layer having a thickness of δ m, the pressure difference Δp across any hole is approximately:

$$\Delta p = \rho_l \frac{v_c^2}{R} \delta\ \text{N} \cdot \text{m}^{-2}$$

It is implicitly assumed that the liquid column having a height of δ m is under the influence of the radial acceleration v_c^2/R.

When the liquid flows through a hole, the pressure head is converted into kinetic energy:

$$\Delta p = \frac{1}{2C^2} \rho_l v_h^2\ \text{N} \cdot \text{m}^{-2}$$

C is the hole coefficient.

It follows that:

$$v_h = v_c \sqrt{\frac{2C^2 \delta}{R}}\ \text{m} \cdot \text{s}^{-1} \tag{5.4}$$

First, the hole velocity (and thus the capacity) is directly proportional to the circumferential velocity. It is clear that the capacity can be controlled by means of the circumferential speed. Second, at a fixed rotational speed, capacity changes can be coped with by δ variations. A rotating device possesses a certain degree of self-control.

Example

The following data apply for Akzo Nobel Base Chemicals' prilling tower for caustic soda (more data in Sec. 5.9):

- $\phi_p = 3000\ \mathrm{kg \cdot hr^{-1}}$
- $\rho_l = 1786\ \mathrm{kg \cdot m^{-3}}$ (at 320°C)
- $D = 0.1699$ m
- $H = 0.1873$ m
- 6000 holes having a diameter of 0.016 in. (0.406 mm)
- $n = 7.5\ \mathrm{sec^{-1}}$

$v_c = 4.00\ \mathrm{m \cdot sec^{-1}}$ and $v_h = 0.61\ \mathrm{m \cdot sec^{-1}}$. It follows that

$$\sqrt{\frac{2C^2\delta}{R}} = 0.1525$$

Furthermore, the radial acceleration is $v_c^2/R = 188.3\ \mathrm{m \cdot sec^{-2}}$, $Fr = 188.3/9.81 = 19.2$.

In other words, the radial acceleration is approximately 20 times greater than the acceleration due to gravity. Under such conditions, the melt is forced against the wall.

Second, the size of the droplets from a rotating device will be dealt with. Van den Berg and Hallie reported that, the load being the same, the diameter of the droplets is considerably smaller than the diameter of the droplets coming from a stationary nozzle. The authors prilled a calcium nitrate solution at 140°C by means of small prilling cups.

The prills coming from the Akzo Nobel Base Chemicals tower previously referred to had a weight average particle size of 0.71 mm. The particle size distribution was narrow. The weight average particle size was not strongly dependent on the rotational speed. In this case, $d_p = 1.75 \cdot d_h$. Wells and Kern (1979) reported that urea prills were obtained by means of a rotating basket. The basket rotated with a speed of 5 $\mathrm{sec^{-1}}$ and contained orifices from 0.95 to 1.2 mm in diameter. The weight average particle size was 1.55 mm, while the particle size distribution was narrow. Here d_p is approximately 1.5 times the hole size. The prilling operation described by Wells and Kern will be considered in more detail in Sec. 5.9.

It was shown that the capacity of a rotating device is strongly dependent on the rotational speed. Furthermore, a spinning cup can adjust to feed rate changes at a certain rotational speed. It was also explained that the average particle size is neither affected strongly by the rotational speed nor by feed rate changes. A tentative conclusion is that the diameter of the droplet is approximately 1.5–2 times the orifice diameter if the velocity is in the range 1–10 $m \cdot sec^{-1}$ and the viscosity is in the range 10^{-3}–10^{-1} $N \cdot sec \cdot m^{-2}$.

The horizontal distance reached by the prills is a function of the drop size and the initial velocity of the drop. According to Eq. (5.4), the initial velocity is directly proportional to the circumferential velocity. Thus the drops leaving a truncated cone (see Fig. 5.4) have different velocities. The maximum diameter of the bucket can be twice the minimum diameter. This construction enables an effective use of the tower. It is also possible to vary the hole size from the top to the bottom of the bucket.

5.3.4. Two-Fluid Nozzles

Two-fluid spray nozzles are used to generate fine droplets. Atomizing energy is supplied by high relative gas velocity.

5.4. TERMINAL VELOCITY OF A FALLING SPHERE

The terminal velocity follows from the equation:

$$\frac{\pi}{6} d_p^3 (\rho_s - \rho_g) g = c_w \frac{\pi}{4} d_p^2 \frac{1}{2} \rho_g v_p^2 \tag{5.5}$$

If $Re \leq 1$, then $c_w = \dfrac{24}{Re}$ and

$$v_p = \frac{(\rho_s - \rho_g) g}{18 \mu_g} d_p^2 \ \text{m} \cdot \text{sec}^{-1} \ \text{(Stokes' Law)}$$

If $10^3 < Re < 10^5$, then $c_w = 0.43$ and

$$v_p = 1.76 \sqrt{\frac{(\rho_s - \rho_g) g}{\rho_g} d_p} \ \text{m} \cdot \text{sec}^{-1}$$

If $1 < Re < 10^3$, Fig. 5.7 should be used.

A terminal velocity v_p is assumed. Table 5.3 can be used to arrive at an assumption that makes sense. The next step is the calculation of $Re = \rho_g v_p d_p / \mu_g$. A c_w-value is read from Fig. 5.7. Now the left-hand side of Eq. (5.5) should be equal to the right-hand side. Otherwise, the aforementioned approach should be repeated.

It can be proven that the particles do not interact on settling.

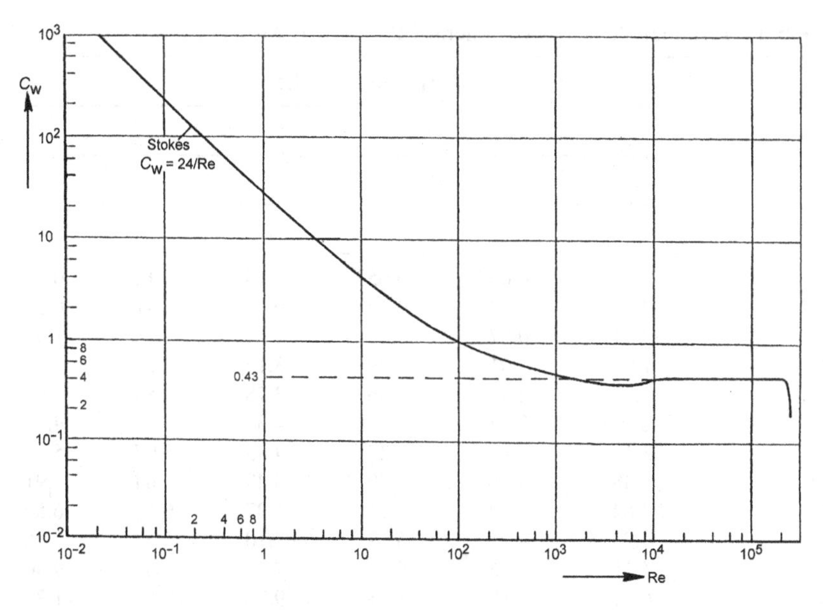

FIG. 5.7 $c_w = f(Re)$. (From Smith et al., 1991.)

5.5. HEAT TRANSFER FROM A FALLING SPHERE

The heat transfer coefficient α_0 is calculated by using Frössling's equation (Frössling, 1938).

$$Nu = 2 + 0.552\ Re^{1/2} Pr^{1/3} \tag{5.6}$$

The evaporation of droplets of nitrobenzene, aniline, and water was investigated. The results can be used for the prediction of heat transfer coefficients due to the analogy between heat transfer and mass transfer.

Droplet sizes: 0.1–0.9 mm
Air velocities: 0.2–7 $m \cdot sec^{-1}$
Re: 2–800

For the sublimation of naphthalene spheres, the same result was obtained with the temperature of the experiments being 20°C.

5.6. SOLIDIFICATION OF A SPHERE

The upper part of Fig. 5.8 depicts the solidification process according to Boretzky (1967). The melt and the solidified shell are both at the melting

TABLE 5.3 Terminal Velocities of Spherical Particles in Air

	$\Delta\rho=(\rho_s-\rho_g)$ (kg/m³)					
	1000			2000		
	(°C)			(°C)		
Diameter (μm)	15	90	205	15	90	205
50	0.074	0.062	0.052	0.14	0.12	0.10
100	0.25	0.22	0.19	0.46	0.41	0.37
250	0.91	0.87	0.83	1.51	1.48	1.43
500	1.98	2.01	1.99	3.17	3.20	3.26
600	2.39	2.43	2.46	3.78	3.84	3.94
700	2.79	2.87	2.91	4.33	4.51	4.70
800	3.14	3.23	3.35	4.85	5.07	5.27
900	3.48	3.66	3.79	5.42	5.64	5.91
1000	3.84	4.00	4.21	5.97	6.19	6.50
1250	4.73	4.91	5.12	7.10	7.53	7.86
1500	5.37	5.76	5.98	7.78	8.65	9.40
2000	6.71	7.18	7.71	9.72	10.6	11.7
2500	7.71	8.45	9.20	11.3	12.5	13.7
3000	8.75	9.56	10.5	14.8	14.0	15.5
5000	11.6	13.1	14.6	16.7	18.6	21.2

Source: Nonhebel and Moss (1971).

point. This case and the next case will be treated mathematically. It is convenient to define three dimensionless numbers at this stage.

$$Bi = \frac{R/\lambda_s}{1/\alpha_o} = \frac{\alpha_o R}{\lambda_s} \quad \text{(Biot number)}$$

The Biot number divides the internal resistance to heat transfer by the external resistance to heat transfer.

$$Ph = \frac{i}{c_s(T_o - T_c)} \quad \text{(Phase Transfer number)}$$

The Phase Transfer number divides the latent heat (the heat of crystallization) by sensible heat.

$$Fo = \frac{at}{R^2} \quad \text{(Fourier number)}$$

The Fourier number is used in many problems concerning instationary heat transfer; the parameter a equals $\lambda_s/(c_s\rho_s)$.

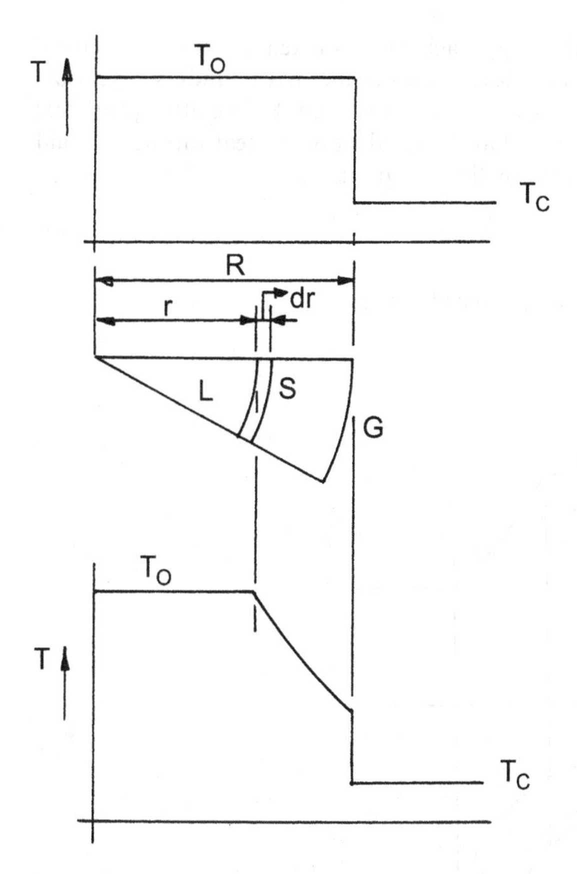

FIG. 5.8 Temperature profiles for a solidifying sphere.

The differential equation for the case depicted in the upper part of Fig. 5.8 is:

$$\alpha_o 4\pi R^2 (T_o - T_c)\mathrm{d}t = -\rho_s i 4\pi r^2 \cdot \mathrm{d}r$$

Integration from $r = R$ to $r = 0$ yields the time for solidification:

$$t_s = \frac{\rho_s i d_p}{6\alpha_o (T_o - T_c)} \text{ sec} \tag{5.7}$$

Dimensionless expression:

$$Fo = \frac{Ph}{3Bi} \tag{5.8}$$

Stein (1971) described an approach where the temperature drops from T_o to T_c through the solidified layer and the thermal boundary layer surrounding the sphere (see the lower part of Fig. 5.8). Before attempting this approach, it is useful to realize that the conduction of heat through a solid hollow sphere can be described by the equation:

$$Q = \frac{\lambda_s}{R_2 - R_1} 4\pi R_1 R_2 (T_1 - T_2) \text{ W} \tag{5.9}$$

(see Fig. 5.9). This will be explained as follows.

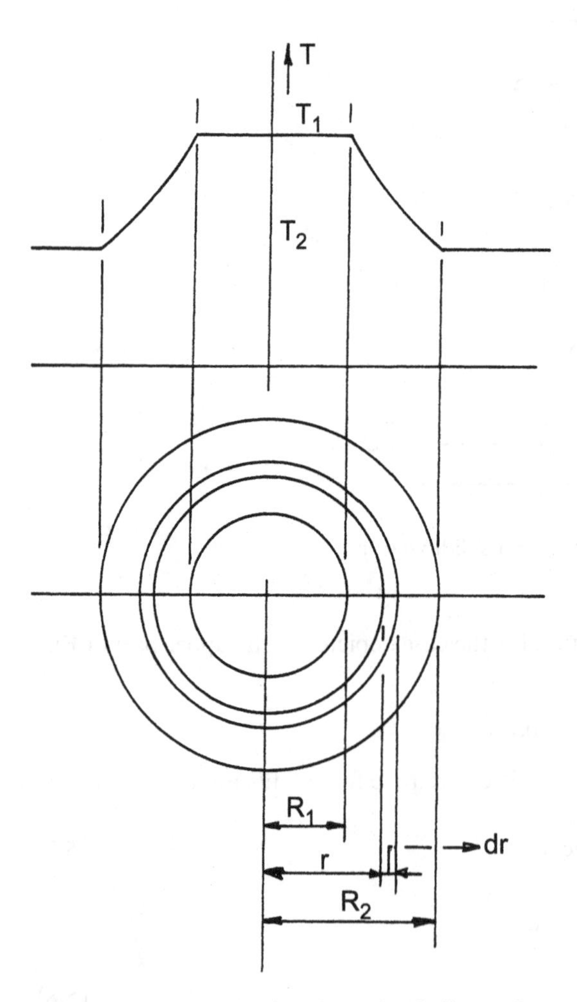

FIG. 5.9 Heat conduction through a hollow sphere.

The heat flow from the inner part of the hollow sphere to the outside is constant. Therefore the heat flow through any thin shell having a thickness dr can be expressed as follows: $Q=-(\lambda_s/\mathrm{d}r)4\pi r^2 \cdot \mathrm{d}T$ W. dT stands for the temperature drop across the thin shell. Integrating the differential equation yields Eq. (5.9).

In the latter equation, the area for heat transfer is the geometric average of the inner and outer area while the coefficient for heat transfer is $\lambda_s/(R_2-R_1)$. The differential equation describing the case depicted in the lower part of Fig. 5.8 is:

$$k4\pi r^2(T_o - T_c)\mathrm{d}t = -\rho_s i 4\pi r^2 \cdot \mathrm{d}r$$

This equation neglects the enthalpy effect of the cooling sphere. The enthalpy effect will be discussed after the completion of the treatment of the present case. The aforementioned differential equation assumes the melt to be at the melting point T_o. If the melt is at a temperature T_f, it is recommended to replace i by $i + c_l(T_f - T_o)$. For this approximation, $(T_f - T_o)$ should be in the range 5–20 K.

k is defined by the following equation:

$$\frac{1}{kr^2} = \frac{1}{\alpha_o R^2} + \frac{R-r}{\lambda_s R r}$$

Integration of the differential equation between $r=R$ and $r=0$ yields the time for solidification:

$$t_s = \frac{\rho_s i d_p}{6\alpha_o(T_o - T_c)}\left(1 + \frac{\alpha_o d_p}{4\lambda_s}\right) \text{ sec} \tag{5.10}$$

Dimensionless expression:

$$Fo = Ph\left(\frac{1}{6} + \frac{1}{3Bi}\right) \tag{5.11}$$

Equations (5.10) and (5.11) were derived by neglecting the specific heat of the solid material. The equations tend to become complicated when this effect is taken into account (see, e.g., Stein, 1971). VDI Heat Atlas (1993) contains a graph that can be used in conjunction with Eq. (5.11). This graph is reproduced in Fig. 5.10. First, Fo is calculated by means of Eq. (5.11). $\tau_{E,min}$ is identical to Fo. Second, a correction factor can be read on the ordinate. This correction factor $\tau_E/\tau_{E,min}$ is a function of both Ph and Bi. VDI Heat Atlas defines Ph in a slightly different way: $Ph = (\rho_l/\rho_s) \cdot i/\{c_s(T_o - T_c)\}$. It is recommended, however, to use the definition for Ph as given in this section.

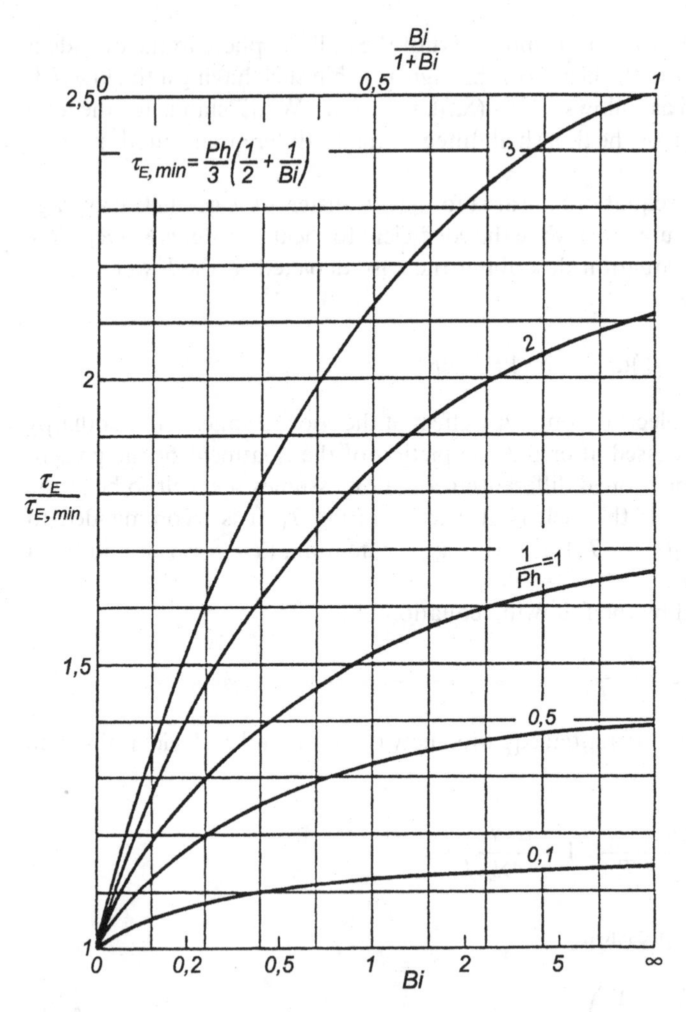

FIG. 5.10 Correction factor for the solidification time. (From VDI, 1993.)

Fig. 5.10 is based on numerical calculations by Tao (1967). Usually, for prilling, the correction factor is in the range 1.2–1.4. This can be explained as follows. A normal value for Ph is $Ph = i/\{c_s(T_o - T_c)\} = 200{,}000/\{2000(120-30)\} = 1.11$ and $1/Ph = 0.9$. A normal value for Bi when inorganic materials are prilled is $Bi = \alpha_o R/\lambda_s = 300 \cdot 0.5 \cdot 10^{-3}/1 = 0.15$. On prilling organic materials, $Bi = 300 \cdot 0.5 \cdot 10^{-3}/0.2 = 0.75$. It will be clear from Fig. 5.10 that correction factors in the range 1.2–1.4 will be obtained.

5.7. PRILL COOLING

A simple approach is suggested.

$$Q_c = \phi_p c_s (T_o - \bar{T}) \text{ W}$$

This is the heat flow for cooling. It is assumed that the prills are, on entering the cooling zone, at the melting point.

$$Q_c = k_c F \Delta T_m \text{ W}$$

This is the heat flow that can be carried away. The heat flow that can be carried away is made equal to the heat flow that should be carried away. The heat transfer coefficient k_c is defined as follows:

$$\frac{1}{k_c R^2} = \frac{1}{\alpha_o R^2} + \frac{2(R - R/2)}{\lambda_s R^2}$$

This approach assumes that the heat is concentrated halfway between $r = R$ and $r = 0$. The heat has to diffuse through a hollow sphere having a shell thickness of $R/2$. Next, the heat is absorbed by the cooling air.

$$\Delta T_m = \frac{(T_o - T_c) - (\bar{T} - T_c)}{{}^e\log \dfrac{T_o - T_c}{\bar{T} - T_c}} \text{ K}$$

It follows that:

$$\phi_p c_s (T_o - \bar{T}) = k_c F \frac{(T_o - T_c) - (\bar{T} - T_c)}{{}^e\log \dfrac{T_o - T_c}{\bar{T} - T_c}}$$

and

$$\bar{T} = T_c + \frac{T_o - T_c}{\exp \dfrac{k_c F}{\phi_p c_s}} \text{ K}$$

The latter equation can be rewritten as:

$$\bar{T} = T_c + \frac{T_o - T_c}{\exp \dfrac{6 k_c t_c}{\rho_s c_s d_p}} \tag{5.12}$$

t_c is the residence time in the cooling zone. The residence time in the cooling zone is a function of the particle size.

Rearranging Eq. (5.12) leads to:

$$t_c = \frac{\rho_s c_s d_p}{6k_c} {}^{e}\log \frac{T_o - T_c}{\bar{T} - T_c} \text{ sec} \tag{5.13}$$

5.8. REVIEW OF THE DESIGN METHOD

The elements of a design method were discussed in the previous sections. Firstly, the terminal velocity of a falling sphere was dealt with. Secondly, the heat transfer from a falling sphere to the air was discussed. Thirdly, the solidification time of a falling sphere was treated, and finally, prill cooling received attention. It is possible to integrate these aspects in order to obtain a design method. The starting points for this approach are:

- The physical properties of the liquid phase are constant.
- The physical properties of the solid phase are constant.
- The air temperature is the arithmetic average of the air inlet and outlet temperatures.
- The volume changes due to different specific masses of liquid and solid are neglected.
- The terminal velocity of a solid sphere is determined.
- The sphere travels through the prill tower at terminal velocity.
- The velocity vector of the sphere has a vertical component only.
- Frössling's equation can be used to describe the heat transfer from a falling sphere to the air.

5.9. VERIFICATION OF THE DESIGN METHODS

5.9.1. Large-Scale Results With Urea

Large-scale results were reported by Wells and Kern (1979).

Urea Physical Properties

T_o = 132.6°C
i = 224,457 $J \cdot kg^{-1}$
ρ_l = 1230 $kg \cdot m^{-3}$ (133°C)
ρ_s = 1335 $kg \cdot m^{-3}$ (20°C)
c_l = 2098 $J \cdot kg^{-1} \cdot K^{-1}$
c_s = 1748 $J \cdot kg^{-1} \cdot K^{-1}$ (25–132°C)
λ_l = 0.83 $W \cdot m^{-1} \cdot K^{-1}$
λ_s = 1.19 $W \cdot m^{-1} \cdot K^{-1}$ (25–89°C)
μ_l = $2.16 \cdot 10^{-3}$ $N \cdot sec \cdot m^{-2}$ (150°C)

Air Physical Properties (36.1 °C)

$\rho_g = 1.14 \text{ kg} \cdot \text{m}^{-3}$
$c_g = 1008 \text{ J} \cdot \text{kg}^{-1} \cdot \text{K}^{-1}$
$\mu_g = 1.90 \cdot 10^{-5} \text{ N} \cdot \text{sec} \cdot \text{m}^{-2}$
$\lambda_g = 0.0268 \text{ W} \cdot \text{m}^{-1} \cdot \text{K}^{-1}$

Equipment

Tower height: 30.33 m
Tower diameter: 15 m
A basket rotates with a speed of 300 min^{-1} and contains orifices from 0.95 to 1.2 mm in diameter.

Independent Variables

$T_f = 140°\text{C}$
$T_{gi} = 26°\text{C}$
$\phi_v = 390{,}000 \text{ m}^3 \cdot \text{hr}^{-1}$ at standard temperature and pressure
$\phi_p = 34{,}380 \text{ kg} \cdot \text{hr}^{-1}$

Dependent Variables

$\overline{T} = 100°\text{C}$

Particle size distribution

Interval (mm)	Mass fraction
0.8–1.0	0.063
1.0–1.2	0.101
1.2–1.4	0.172
1.4–1.6	0.200
1.6–1.8	0.200
1.8–2.0	0.142
2.0–2.2	0.074
2.2–2.4	0.048

The average particle size is somewhat smaller than 1.89 times the diameter of the orifice (see Sec. 5.3). Furthermore, there is a particle size distribution instead of a uniform particle size.

Calculations

The calculations will be carried out for prills having a diameter of 1.5 mm.

1. *Heat balance*

$$34,380\{2.098(140-132.6)+224.457+1.748(132.6-100)\}$$
$$=34,380\cdot 296.967=10,209,725\ \text{kJ}\cdot\text{hr}^{-1}=2836\ \text{kW}$$

2. *ΔT airflow*

$$\Delta T=\frac{10,209,725}{1.008\cdot 390,000\cdot 1.29}=20.1\ \text{K}$$

3. *Average air temperature* (°C)

$$26+0.5\cdot 20.1=36.1°\text{C}$$

4. *Average air velocity*

$$\frac{390,000\cdot 4}{\pi\cdot 225\cdot 3600}\cdot\frac{1.29}{1.14}=0.69\ \text{m}\cdot\text{sec}^{-1}$$

5. *Terminal velocity*

$$d_p=1.5\cdot 10^{-3}\ \text{m}$$

Assume $v_r=6.3\ \text{m}\cdot\text{sec}^{-1}$

$$Re=\frac{1.14\cdot 6.3\cdot 1.5\cdot 10^{-3}}{1.90\cdot 10^{-5}}=567$$

$c_w=0.57$ (see Fig. 5.7)

Equation (5.5):

$$\frac{\pi}{6}d_p^3(\rho_s-\rho_g)g\overset{?}{=}c_w\frac{\pi}{4}d_p^2\cdot\frac{1}{2}\rho_g v_r^2$$

$$\frac{\pi}{6}(1.5\cdot 10^{-3})^3(1335-1.14)9.81=2.31\cdot 10^{-5}$$

$$0.57\cdot\frac{\pi}{4}(1.5\cdot 10^{-3})^2\frac{1}{2}\cdot 1.14\cdot 6.3^2=2.28\cdot 10^{-5}$$

$v_r=6.3\ \text{m}\cdot\text{sec}^{-1}$
$v_a=6.3-0.7=5.6\ \text{m}\cdot\text{sec}^{-1}$

6. *Heat transfer coefficient*

$$Re=567$$

$$Pr=\frac{\mu_g c_g}{\lambda_g}=\frac{1.90\cdot 10^{-5}\cdot 1008}{0.0268}=0.71$$

$$Nu=2+0.552\cdot 567^{1/2}\cdot 0.71^{1/3}=13.72$$

$$\frac{\alpha_o d_p}{\lambda_g} = 13.72$$

$$\alpha_o = \frac{\lambda_g}{d_p} \cdot 13.72 = \frac{0.0268}{1.5 \cdot 10^{-3}} \cdot 13.72 = 245.1 \text{ W} \cdot \text{m}^{-2} \cdot \text{K}^{-1}$$

7. *Solidification time*

$$Ph = \frac{i + c_l(T_f - T_o)}{c_s(T_o - T_c)} = \frac{224,457 + 2098(140 - 132.6)}{1748(132.6 - 36.1)} = 1.42$$

$$Bi = \frac{\alpha_o d_p}{2\lambda_s} = \frac{245.1 \cdot 1.5 \cdot 10^{-3}}{2 \cdot 1.19} = 0.154$$

$$Fo = Ph\left(\frac{1}{6} + \frac{1}{3Bi}\right) = 1.42\left(\frac{1}{6} + \frac{1}{3 \cdot 0.154}\right) = 3.31$$

$$a = \frac{\lambda_s}{c_s \rho_s} = \frac{1.19}{1748 \cdot 1335} = 5.10 \cdot 10^{-7}$$

$$t_{s1} = \frac{Fo \cdot d_p^2}{4a} = \frac{3.31(1.5 \cdot 10^{-3})^2}{4 \cdot 5.10 \cdot 10^{-7}} = 3.7 \text{ sec}$$

$\tau_E/\tau_{E,min} = 1.18$ (see Fig. 5.10)

$t_{s2} = 1.18 \cdot 3.7 = 4.4$ sec

8. *Tower height for solidification*

$4.4 \cdot 5.6 = 24.6$ m

9. *Prill cooling*

$$R = \frac{d_p}{2} = 7.5 \cdot 10^{-4} \text{ m}$$

$$\frac{1}{\alpha_o R^2} = 7253 \text{ W}^{-1} \cdot \text{K}$$

$$\frac{2(R - R/2)}{\lambda_s R^2} = \frac{2(7.5 \cdot 10^{-4} - 3.75 \cdot 10^{-4})}{1.19 \cdot 7.5^2 \cdot 10^{-8}} = 1120$$

$$\frac{1}{k_c R^2} = 7253 + 1120 = 8373$$

$$k_c = \frac{1}{8373 \cdot 7.5^2 \cdot 10^{-8}} = 212.3 \text{ W} \cdot \text{m}^{-2} \cdot \text{K}^{-1}$$

Available cooling height: $30.33 - 24.6 = 5.7$ m.

$$t_c = \frac{5.7}{5.6} = 1.0 \text{ sec}$$

$$\overline{T} = T_c + \frac{T_o - T_c}{\exp \dfrac{6k_c t_c}{\rho_s c_s d_p}} = 36.1 + \frac{132.6 - 36.1}{\exp \dfrac{6 \cdot 212.3 \cdot 1.0}{1335 \cdot 1748 \cdot 1.5 \cdot 10^{-3}}}$$

$$= 103.2°C$$

10. *Comments*

It can be seen that the model predicts the solidification and cooling of prills having a diameter of 1.5 mm reasonably well. Smaller prills will leave the tower at a lower temperature, while larger prills can leave the tower partly solidified. The latter aspect is not a problem as the outer shell will be solid. Ultimately, during transport and storage, the bulk material will attain a temperature of 100–110°C.

5.9.2. Large-Scale Results with Caustic Soda

Large-scale results were obtained from an Akzo Nobel prill tower.

Caustic Soda Physical Properties

T_o = 318.4°C
i = 270,700 $J \cdot kg^{-1}$ (includes the transition NaOH-α to NaOH-β at 299.6°C)
ρ_l = 1786 $kg \cdot m^{-3}$ (320°C)
ρ_s = 2055 $kg \cdot m^{-3}$ (230°C)
c_l = 2150 $J \cdot kg^{-1} \cdot K^{-1}$ (330°C)
c_s = 1950 $J \cdot kg^{-1} \cdot K^{-1}$ (230°C)
λ_l = 0.84 $W \cdot m^{-1} \cdot K^{-1}$ (330°C)
λ_s = 1.6 $W \cdot m^{-1} \cdot K^{-1}$ (230°C)
μ_l = $3.6 \cdot 10^{-3}$ $N \cdot sec \cdot m^{-2}$ (360°C)

Note

Data regarding the thermal conductivity of solid NaOH could not be found. A comparative study carried out by Hittenhausen (1994) showed that the conductivity of solid NaOH will be equal to the conductivity of solid LiOH. Janz et al. (1979) give measured data regarding the conductivity of solid LiOH. The melting point of LiOH is 462 ± 5°C and its thermal conductivity at 230°C is 1.6 $W \cdot m^{-1} \cdot K^{-1}$.

Air Physical Properties (138.5°C)

$$\rho_g = 0.856 \text{ kg} \cdot \text{m}^{-3}$$
$$c_g = 1019 \text{ J} \cdot \text{kg}^{-1} \cdot \text{K}^{-1}$$
$$\mu_g = 2.35 \cdot 10^{-5} \text{ N} \cdot \text{sec} \cdot \text{m}^{-2}$$
$$\lambda_g = 0.034 \text{ W} \cdot \text{m}^{-1} \cdot \text{K}^{-1}$$

Equipment

Tower height: 22.2 m
Tower diameter: 8.75 m

A basket rotates with a speed of 450 min^{-1} and contains orifices having a diameter of 0.016 in. (0.406 mm). The basket diameter is 170 mm.

Independent Variables

$$T_f = 341°\text{C}$$
$$T_{gi} = 129°\text{C}$$
$$\phi_p = 3.0 \text{ t} \cdot \text{hr}^{-1}$$

Dependent Variables

$$\overline{T} = 145°\text{C}$$
$$T_{go} = 148°\text{C}$$

Particle size distribution

Interval (mm)	Mass fraction
<0.50	0.0234
0.50–0.60	0.0973
0.60–0.71	0.3375
0.71–0.85	0.3578
0.85–1.19	0.1643
1.19–1.41	0.0171
>1.41	0.0025

The weight average particle size (0.73 mm) is close to $1.89 \cdot 0.406 = 0.77$ mm (see Sec. 5.3). The peripheral velocity of the rotating basket is $4.0\ \mathrm{m \cdot sec^{-1}}$. The particle size distribution is narrow.

Calculations

The calculations will be carried out for particles having a diameter of 0.78 mm.

1. *Heat balance*

$$3000\{2.15(341 - 318.4) + 270.7 + 1.95(318.4 - 145)\}$$

$$= 1{,}972{,}260\ \mathrm{kJ \cdot hr^{-1}} = 548\ \mathrm{kW}$$

2. *Airflow*

$$\phi_m = \frac{1{,}972{,}260}{1.019(148 - 129)} = 101{,}868\ \mathrm{kg \cdot hr^{-1}}$$

3. *Average air temperature (°C)*

$(129 + 148)/2 = 138.5°\mathrm{C}$

4. *Average air velocity*

$$\frac{4 \cdot 101{,}868}{\pi \cdot 8.75^2 \cdot 3600 \cdot 0.856} = 0.55\ \mathrm{m \cdot sec^{-1}}$$

5. *Terminal velocity*

$d_p = 7.8 \cdot 10^{-4}$ m

Assume $v_r = 5.6\ \mathrm{m \cdot sec^{-1}}$

$$Re = \frac{0.856 \cdot 5.6 \cdot 7.8 \cdot 10^{-4}}{2.35 \cdot 10^{-5}} = 159.1$$

$c_w = 0.8$ (see Fig. 5.7).

Equation (5.5):

$$\frac{\pi}{6} d_p^3 (\rho_s - \rho_g) g \stackrel{?}{=} c_w \frac{\pi}{4} d_p^2 \cdot \frac{1}{2} \rho_g v_r^2$$

$$\frac{\pi}{6} (0.78 \cdot 10^{-3})^3 (2055 - 0.856) 9.81 = 5.01 \cdot 10^{-6}$$

$$0.8 \cdot \frac{\pi}{4} (7.8 \cdot 10^{-4})^2 \frac{1}{2} \cdot 0.856 \cdot 5.6^2 = 5.13 \cdot 10^{-6}$$

$v_a = 5.6 - 0.55 \approx 5.0\ \mathrm{m \cdot sec^{-1}}$

6. *Heat transfer coefficient*

$Re = 159.1$

$$Pr = \frac{\mu_g c_g}{\lambda_g} - \frac{2.35 \cdot 10^{-5} \cdot 1019}{0.034} = 0.70$$

$$Nu = 2 + 0.552 \cdot 159.1^{1/2} \cdot 0.70^{1/3} = 8.18$$

$$\frac{\alpha_o d_p}{\lambda_g} = 8.18$$

$$\alpha_o = \frac{\lambda_g}{d_p} \cdot 8.18 = \frac{0.034}{7.8 \cdot 10^{-4}} \cdot 8.18 = 356.6 \text{ W} \cdot \text{m}^{-2} \cdot \text{K}^{-1}$$

7. *Solidification time*

$$Ph = \frac{i + c_l(T_f - T_o)}{c_s(T_o - T_c)} = \frac{270,700 + 2150(341 - 318.4)}{1950(318.4 - 138.5)} = 0.91$$

$$Bi = \frac{\alpha_o d_p}{2\lambda_s} = \frac{356.6 \cdot 7.8 \cdot 10^{-4}}{2 \cdot 1.6} = 0.087$$

$$Fo = Ph\left(\frac{1}{6} + \frac{1}{3Bi}\right) = 0.91\left(\frac{1}{6} + \frac{1}{3 \cdot 0.087}\right) = 3.64$$

$$a = \frac{\lambda_s}{c_s \rho_s} = \frac{1.6}{1950 \cdot 2055} = 3.99 \cdot 10^{-7}$$

$$t_{sl} = \frac{Fo \cdot d_p^2}{4a} = \frac{3.64(7.8 \cdot 10^{-4})^2}{4 \cdot 3.99 \cdot 10^{-7}} = 1.4 \text{ sec}$$

$\tau_E/\tau_{E,min} = 1.18$ (see Fig. 5.10)

$t_{s2} = 1.18 \cdot 1.4 = 1.7$ sec

8. *Tower height for solidification*

$1.7 \cdot 5.0 = 8.5$ m

9. *Prill cooling*

$R = d_p/2 = 3.9 \cdot 10^{-4}$ m

$$\frac{1}{\alpha_o R^2} = 18{,}437 \text{ W}^{-1} \cdot \text{K}$$

$$\frac{2(R-R/2)}{\lambda_s R^2} = \frac{2(3.9 \cdot 10^{-4} - 1.95 \cdot 10^{-4})}{1.6 \cdot 3.9^2 \cdot 10^{-8}} = 1603$$

$$\frac{1}{k_c R^2} = 18{,}437 + 1603 = 20{,}040$$

$$k_c = \frac{1}{20{,}040 \cdot 3.9^2 \cdot 10^{-8}} = 328.1 \ \mathrm{W \cdot m^{-2} \cdot K^{-1}}$$

Available cooling height: 22.2 – 8.5 = 13.7 m

$$t_c = \frac{13.7}{5.0} = 2.7 \ \mathrm{sec}$$

$$\overline{T} = T_c + \frac{(T_o - T_c)}{\exp \dfrac{6k_c t_c}{\rho_s c_s d_p}} = 138.5 + \frac{318.4 - 138.5}{\exp \dfrac{6 \cdot 328.1 \cdot 2.7}{2055 \cdot 1950 \cdot 7.8 \cdot 10^{-4}}}$$

$$= 171.3 ^\circ\mathrm{C}$$

10. *Comments*

It can be seen that the model predicts the solidification and cooling of prills having a diameter of 0.78 mm reasonably well. The prediction regarding the cooling of the prills seems high. The prilled material leaves the tower at 145°C. The prediction for particles having a diameter of 0.78 mm is 171.3°C. First, the weight average particle size is 0.73 mm. The prediction for particles having a diameter of 0.73 mm will be lower than 171.3°C. Second, the prediction will always be slightly too high because a constant air temperature of 138.5°C has been taken for the calculations.

11. *Additional remarks*

Caustic soda is prilled in a relatively warm airflow. The cooling air enters at 129°C. The reason is that caustic soda is very hygroscopic at ambient temperature. At temperatures well above 100°C, water pickup cannot occur.

At 3 $\mathrm{t \cdot hr^{-1}}$, the prill tower ran at approximately 50% of the nominal capacity. Prilling at higher throughputs is possible; however, the air outlet temperature rises to temperatures as high as 170–180°C. Taking the arithmetic average of the air inlet and air outlet temperatures is then an approximation only. The prediction of the product exit temperature becomes less accurate.

5.10. DESIGN EXAMPLE

It is desired to prill 5000 kg of stearic acid per hour. A particle size of 1.2 mm is required. The melt temperature is 80°C. The cooling air enters the tower at

15°C and leaves at 25°C. The solid stearic acid particles should be cooled down to 30°C.

Stearic Acid Physical Properties

T_o = 70.0°C
i = 215,161 J · kg^{-1}
ρ_l = 839 kg · m^{-3} (80°C)
ρ_s = 1009 kg · m^{-3} (50°C)
c_l = 2,306 J · kg^{-1} · K^{-1} (77°C)
c_s = 2077 J · kg^{-1} · K^{-1} (47°C)
λ_l = 0.170 W · m^{-1} · K^{-1} (80°C)
λ_s = 0.175 W · m^{-1} · K^{-1} (Boretzky, 1967)
μ_l = 7.7·10^{-3} N · sec · m^{-2} (80°C)

Air Physical Properties (20°C)

ρ_g = 1.202 kg · m^{-3}
c_g = 1007 J · kg^{-1} · K^{-1}
μ_g = 1.82 · 10^{-5} N · sec · m^{-2}
λ_g = 0.0257 W · m^{-1} · K^{-1}

1. *Heat balance*

$$5000\{2.306(80 - 70) + 215.161 + 2.077(70 - 30)\}$$
$$= 1,606,505 \text{ kJ} \cdot \text{hr}^{-1} = 446.3 \text{ kW}$$

2. *Airflow*

$$\frac{1,606,505}{1.007 \cdot 10} = 159,534 \text{ kg} \cdot \text{hr}^{-1} = 132,724 \text{ m}^3 \cdot \text{hr}^{-1}$$

3. *Tower diameter*

An upward air superficial velocity of 0.5 m · sec^{-1} is taken.

$$D = \sqrt{\frac{132,724 \cdot 4}{\pi \cdot 0.5 \cdot 3600}} = 9.69 \text{ m}$$

Take a tower diameter of 10 m leading to an air superficial velocity of 0.47 m · sec^{-1}.

4. *Terminal velocity*

$d_p = 1.2 \cdot 10^{-3}$ m

Assume $v_r = 4.5 \cdot \text{m} \cdot \text{sec}^{-1}$

$$Re = \frac{1.202 \cdot 4.5 \cdot 1.2 \cdot 10^{-3}}{1.82 \cdot 10^{-5}} = 356.6$$

$c_w = 0.65$ (see Fig. 5.7).

Equation (5.5):

$$\frac{\pi}{6} d_p^3 (\rho_s - \rho_g) g \stackrel{?}{=} c_w \frac{\pi}{4} d_p^2 \cdot \frac{1}{2} \rho_g v_r^2$$

$$\frac{\pi}{6}(1.2 \cdot 10^{-3})^3 (1009 - 1.202) 9.81 = 8.95 \cdot 10^{-6}$$

$$0.65 \cdot \frac{\pi}{4}(1.2 \cdot 10^{-3})^2 \frac{1}{2} \cdot 1.202 \cdot 4.5^2 = 8.95 \cdot 10^{-6}$$

$$v_a = 4.5 - 0.47 = 4.0 \text{ m} \cdot \text{sec}^{-1}$$

5. *Heat transfer coefficient*

$$Re = 356.6$$

$$Pr = \frac{\mu_g c_g}{\lambda_g} = \frac{1.82 \cdot 10^{-5} \cdot 1007}{0.0257} = 0.71$$

$$Nu = 2 + 0.552 \cdot 356.6^{1/2} \cdot 0.71^{1/3} = 11.30$$

$$\frac{\alpha_o d_p}{\lambda_g} = 11.30$$

$$\alpha_o = \frac{\lambda_g}{d_p} \cdot 11.30 = \frac{0.0257}{1.2 \cdot 10^{-3}} \cdot 11.30 = 242.0 \text{ W} \cdot \text{m}^{-2} \cdot \text{K}^{-1}$$

6. *Solidification time*

$$Ph = \frac{i + c_l(T_f - T_o)}{c_s(T_o - T_c)} = \frac{215,161 + 2306(80 - 70)}{2077(70 - 20)} = 2.29$$

$$Bi = \frac{\alpha_o d_p}{2\lambda_s} = \frac{242.0 \cdot 1.2 \cdot 10^{-3}}{2 \cdot 0.175} = 0.830$$

$$Fo = Ph\left(\frac{1}{6} + \frac{1}{3Bi}\right) = 2.29\left(\frac{1}{6} + \frac{1}{3 \cdot 0.830}\right) = 1.30$$

$$a = \frac{\lambda_s}{c_s \rho_s} = \frac{0.175}{2077 \cdot 1009} = 8.35 \cdot 10^{-8}$$

$$t_{s1} = \frac{Fo \cdot d_p^2}{4a} = \frac{1.30(1.2 \cdot 10^{-3})^2}{4 \cdot 8.35 \cdot 10^{-8}} = 5.6 \text{ sec}$$

$\tau_E/\tau_{E,min} = 1.28$ (see Fig. 5.10)

$$t_{s2} = 1.28 \cdot 5.6 = 7.2 \text{ sec}$$

7. *Tower height for solidification*

   $7.2 \cdot 4.0 = 28.8$ m

8. *Prill cooling time*

   $R = d_p/2 = 6.0 \cdot 10^{-4}$ m

$$\frac{1}{\alpha_o R^2} = 11,478 \text{ W}^{-1} \cdot \text{K}$$

$$\frac{2(R - R/2)}{\lambda_s R^2} = \frac{2(6.0 \cdot 10^{-4} - 3.0 \cdot 10^{-4})}{0.175 \cdot 6.0^2 \cdot 10^{-8}} = 9524$$

$$\frac{1}{k_c R^2} = 11,478 + 9524 = 21,002$$

$$k_c = \frac{1}{21,002 \cdot 6.0^2 \cdot 10^{-8}} = 132.3 \text{ W} \cdot \text{m}^{-2} \cdot \text{K}^{-1}$$

$$t_c = \frac{\rho_s c_s d_p}{6 k_c} \, {}^{e}\log \frac{T_o - T_c}{\overline{T} - T_c}$$

$$= \frac{1009 \cdot 2077 \cdot 1.2 \cdot 10^{-3}}{6 \cdot 132.3} \cdot {}^{e}\log \frac{70 - 20}{30 - 20} = 5.1 \text{ sec}$$

9. *Prill cooling height*

   $5.1 \cdot 4.0 = 20.4$ m

10. *Total tower height*

    $28.8 + 20.4 = 49.2$ m, take 50 m

11. *Comments*

The tower becomes quite high because the particle size is rather large and because the thermal conductivity of solid stearic acid is low.

LIST OF SYMBOLS

a	Thermal diffusivity $[\lambda_s/(c_s \rho_s)]$ $[\text{m}^2 \cdot \text{sec}^{-1}]$
Bi	Biot number $(\alpha_o R/\lambda_s)$
C	Hole coefficient
c_g	Gas specific heat $[\text{J} \cdot \text{kg}^{-1} \cdot \text{K}^{-1}]$
c_l	Melt specific heat $[\text{J} \cdot \text{kg}^{-1} \cdot \text{K}^{-1}]$
c_s	Solid specific heat $[\text{J} \cdot \text{kg}^{-1} \cdot \text{K}^{-1}]$
c_w	Resistance coefficient (falling particle)
D	Prill tower diameter [m]
	Cup diameter [m]

d_h	Hole diameter [m]
d_p	Particle diameter [m]
F	Heat transfer area [m^2]
Fo	Fourier number (at/R^2)
Fr	Froude number $\{v_c^2/(Rg)\}$
g	Acceleration due to gravity [m · sec^{-2}]
H	Cup height [m]
i	Heat of fusion [J · kg^{-1}]
k	Heat transfer coefficient [W · m^{-2} · K^{-1}]
k_c	Overall heat transfer coefficient for prill cooling [W · m^{-2} · K^{-1}]
Nu	Nusselt number $(\alpha_o d_p/\lambda_g)$
n	Prilling cup rotational speed [sec^{-1}]
Ph	Phase Transfer number $[i/\{c_s(T_o - T_c)\}]$
Pr	Prandtl number $(\mu_g c_g/\lambda_g)$
Δp	Pressure loss [N · m^{-2}]
Q	Heat flow [W]
Q_c	Heat flow for cooling [W]
R	Sphere radius [m]
	Cup radius [m]
	Material thickness [m]
R_1	Hollow sphere inner radius [m]
R_2	Hollow sphere outer radius [m]
Re	Reynolds number $(\rho_g v_p d_p/\mu_g)$
r	Sphere radius (variable) [m]
T	Temperature [°C]
ΔT	Temperature difference [K]
$\overline{T}$	Average prill temperature [°C]
T_c	Cooling medium temperature [°C]
T_f	Melt temperature [°C]
T_{gi}	Cooling gas inlet temperature [°C]
T_{go}	Cooling gas outlet temperature [°C]
T_o	Melting point [°C]
T_1	Hollow sphere temperature at $r = R_1$ [°C]
T_2	Hollow sphere temperature at $r = R_2$ [°C]
ΔT_m	Logarithmic mean temperature difference [K]
t	Time [sec]
t_c	Time for cooling [sec]
t_s	Time for solidification [sec]
t_{s1}	Uncorrected solidification time [sec]
t_{s2}	Corrected solidification time [sec]
v_a	Absolute velocity [m · sec^{-1}]
v_c	Circumferential velocity [m · sec^{-1}]
v_h	Hole velocity [m · sec^{-1}]

v_p	Particle terminal velocity [$m \cdot sec^{-1}$]
v_r	Relative velocity [$m \cdot sec^{-1}$]
α_o	Heat transfer coefficient (gas/prills) [$W \cdot m^{-2} \cdot K^{-1}$]
δ	Liquid layer thickness [m]
λ	Wavelength [m]
λ_g	Gas thermal conductivity [$W \cdot m^{-1} \cdot K^{-1}$]
λ_l	Melt thermal conductivity [$W \cdot m^{-1} \cdot K^{-1}$]
λ_s	Solid thermal conductivity [$W \cdot m^{-1} \cdot K^{-1}$]
μ_g	Gas dynamic viscosity [$N \cdot sec \cdot m^{-2}$]
μ_l	Melt dynamic viscosity [$N \cdot sec \cdot m^{-2}$]
v_l	Melt kinematic viscosity [$m^2 \cdot sec$]
ρ_g	Gas specific mass [$kg \cdot m^{-3}$]
ρ_l	Melt specific mass [$kg \cdot m^{-3}$]
ρ_s	Solid specific mass [$kg \cdot m^{-3}$]
σ	Surface tension [$N \cdot m^{-1}$]
τ_E	Corrected dimensionless solidification time
$\tau_{E,min}$	Uncorrected dimensionless solidification time
ϕ_p	Production [$kg \cdot hr^{-1}$]
ϕ_m	Cooling medium (gas) flow [$kg \cdot hr^{-1}$]
ϕ_v	Cooling medium (gas) flow [$m^3 \cdot hr^{-1}$]

REFERENCES

Boretzky, W. (1967). Spray granulation—a new process to obtain uniform solid spheres from melts. *Fette-Seifen-Anstrichmittel* 69:263. In German.

Duffie, J. A., Marshall, W. R., Jr. (1953). Factors influencing the properties of spray-dried materials. *Chem. Eng. Prog.* 49:417.

Frössling, N. (1938). On the evaporation of falling droplets. *Gerlands Beitr. Geophys. (Leipzig)* 50:170. In German.

Grassmann, P., Reinhart, A. (1961). On the determination of the terminal velocity of droplets and the rising velocity of bubbles. *Chemie-Ingenieur-Technik* 33:348. In German.

Hittenhausen, H. (1994). Thermal conductivity of solid NaOH. *Intern. Rep. Akzo Nobel Eng.* 1.854.489.

Janz, G. J., Allen, C. B., Bansal, N. P., Murphy, R. M., Tomkins, R. P. T. (1979). *Physical Properties Data Compilations Relevant to Energy Storage Part II*. USA: National Bureau of Standards.

Marshall, W. R., Jr. (1954). *Atomization and Spray Drying*. New York: American Institute of Chemical Engineers.

Merrington, A. C., Richardson, E. G. (1947). The break-up of liquid jets. *Proc. Phys. Soc.* 59, 1.

Nonhebel, G., Moss, A. A. H. (1971). *Drying of Solids in the Chemical Industry*. London: Butterworths.

Rayleigh, Lord. (1878). On the instability of jets. *Proc. Lond. Math. Soc.* 10:4.

Ruskan, R. P. (1976). Prilling vs. granulation for nitrogen fertilizer production. *Chem. Eng.* 83:114.
Smith, J. M., Janssen, L. P. B. M., Stammers, E. (1991). *Physical Transport Phenomena I.* Delft, The Netherlands: VSSD. In Dutch.
Stein, W. A. (1971). An approximate solution for the solidification of spherical particles. *Verfahrenstechnik* 5:453. In German.
Tao, L. C. (1967). Generalized numerical solutions of freezing a saturated liquid in cylinders and spheres. *AIChE J.* 13:165.
Van den Berg, P. J., Hallie, G. (1960). *New Developments in Granulation Techniques.* London: The Fertiliser Society.
VDI (1993). *VDI Heat Atlas.* Düsseldorf, Germany: VDI Verlag.
Von Ohnesorge, W. (1936). The formation of droplets at orifices and the break-up of liquid jets. *Z. Angew. Math. Mech.* 16:355. In German.
Weber, C. (1931). On the disintegration of a liquid jet. *Z. Angew. Math. Mech.* 11:136. In German.
Wells, G. L., Kern, J. (1979). A generalised procedure for the design and rating of prilling towers. *Chemie-Ingenieur-Technik*, MS 702/79.

6

The Crystallization of Melts that Tend to Supercool

6.1. INTRODUCTION

Melts that tend to supercool cannot be pastillated or prilled in a straightforward fashion. This is described in Sec. 6.2. The cause of the problem is explained in Sec. 6.3. "Anticrystalline" clusters are present above the melting point. These clusters hinder normal crystallization. This phenomenon occurs in organic melts having viscosities in the range 10^{-2}–1 N · sec · m^{-2} and having relatively complicated molecular structures. For complicated molecular structures, it is not easy to stack the molecules one on top of the other as in crystallization. An abnormal viscosity increase on approaching the melting point downward is an indication for the occurrence of "anticrystalline" clusters. This is treated in Sec. 6.4 and a number of examples are given. Sec. 6.5 deals with equipment for the crystallization of melts that tend to supercool. Pastilles can be made in a Sandvik unit, whereas rods are produced in a twin-screw extruder. Finally, the drown-out of melts to get granules is described. Examples of pastille production and rod production are given in Secs. 6.6 through 6.8.

6.2. THE TECHNOLOGICAL PROBLEM

Organic substances of commercial interest having a melting point above ambient temperature are usually sold as, for example, slabs, flakes, pastilles, or prills. In many instances, the melts crystallize relatively quickly when they pass the melting point while the heat of fusion is carried away. However, there are cases in which crystallization does not start spontaneously after the melting point has been passed. Sometimes, there is only a small delay. Melts experiencing a small delay cannot be pastillated in a straightforward fashion as the melt first flows out into a round platelet and then sets. Prilling is sometimes not possible either as the residence time in the tower is too short and droplets instead of prills hit the bottom. However, it is usually possible to make slabs or flakes out of these melts.

In other instances, the melt can be cooled 10, 20, or 30 K below the melting point and kept at that temperature for, for example, an hour or even a day. Eventually, the viscous melt will crystallize. However, the point is that these long residence times are, in a chemical plant, highly undesirable. Lowering the temperature further is not a remedy as the increase of the driving force is more than compensated by a viscosity increase. Adding seeds will usually not lead to instantaneous crystallization either. Scratching or scraping is successful in a few cases only. Frequently, this type of organic substance can readily be made to pass into the glassy state on cooling.

6.3. THE CAUSE OF THE PROBLEM

It has been explained in Chapter 1 that organic melts form molecular crystals on solidification. The molecules of organic melts are bound together by relatively weak Van der Waals forces to form molecular crystals. It has been shown that melts exhibiting pronounced supercooling already contain "anti-crystalline" clusters on approaching the melting point downward. Prominent prefreezing anomalies can be caused by these clusters if their concentration becomes appreciable.

In the group of molecular crystals, various physical properties have been looked at in the study of prefreezing anomalies. An upturn of specific heat on approaching the melting point is such a prefreezing anomaly. This probably indicates an enthalpy change resulting from cluster formation.

Melts that tend to supercool show an upturn of viscosity on approaching the melting point downward. Viscosity measurements can be used to demonstrate the existence of prefreezing anomalies. An aspect is that viscosity measurements can be readily extended into the supercooled range of temperatures without risk of spontaneous freezing.

Basically, there are two causes for the formation of "anticrystalline" clusters:

- The viscosity of the melt on approaching the melting point downward is in the range 10^{-2}–$1\ N \cdot sec \cdot m^{-2}$.
- The molecular structure is complicated, stacking the molecules one on top of the other is not straightforward.

6.4. VISCOSITY MEASUREMENTS

6.4.1. Introduction

The Orrick and Erbar (1974) method in Poling et al. (2000) for the estimation of low-temperature viscosity was mentioned in Sec. 1.10. This method uses a group contribution technique to estimate A and B:

$$^{e}\log \frac{\mu}{\rho_{l} M} = A + \frac{B}{T} \tag{1.1}$$

μ: liquid viscosity in cP
ρ_l: liquid specific mass at 20°C in $g \cdot cm^{-3}$
M: molecular weight in $g \cdot gmol^{-1}$
T: temperature in K

A simplified form of Eq. (1.1) is:

$$^{e}\log \mu = A' + \frac{B'}{T} \tag{6.1}$$

This simple form is often referred to as the Andrade equation. For melts that tend to supercool, the viscosity measured on approaching the melting point downward is higher than the viscosity predicted by the Andrade equation. This phenomenon can be explained by postulating that part of the volume is blocked for flow, as a consequence of clusters formed in the melt, in increasing concentration as the temperature decreases to T_o and falls below it. A typical cluster behaves as a colloidal particle suspended in the fluid. Einstein showed that the presence of particles in suspension increases the viscosity above the theoretical value by blocking flow.

If the fraction φ of the total volume occupied by such particles does not exceed approximately 0.3, the ratio of the actual viscosity and the predicted viscosity is given by:

$$\mu_{act}/\mu_{pred} = 1 + 2.5\varphi + 7\varphi^2 + \ldots \tag{6.2}$$

The fraction φ can be calculated from observed deviations from the Andrade equation. In many cases, it is found that a plot $^{e}\log \varphi$ vs. $1/T$ gives a straight line (Ubbelohde, 1978).

6.4.2. Viscosity Anomalies in Polyphenyl Melts

Ubbelohde treats this subject. Strong interaction forces, e.g., dipole forces, are largely absent from these molecules. The molecular configurations, which may either favor or obstruct interlocking, are primarily responsible for cluster formation. The behavior of a series of linear polyphenyls remains practically normal on approaching the freezing point of the melt downward. However, prefreezing anomalies are shown by branched polyphenyls. The remarkable viscosity increase becomes very prominent for molecular structures such as *o*-terphenyl and 1,3,5-tri-α-naphthylbenzene (Figs. 6.1 and 6.2). φ_o at the melting point gives a useful indication of the fraction of volume blocked for flow.

6.4.3. Viscosity Anomalies of Diphenylamines

Amino-diphenylamine is an intermediate for the manufacture of antidegradants for rubber whereas the other two substances in Fig. 6.3 are antidegradants (see also Fig. 6.4). Amino-diphenylamine can be shaped in a straightforward fashion. 4-Isopropyl-amino-diphenylamine has a small tendency to supercool. For example, straightforward pastillation is not possible as the deposited droplet flows out into a thin platelet and a pastille is not obtained. 4-Isohexyl-amino-diphenylamine shows a marked tendency to

Molecule	Shape	T_o	φ_o
o-Terphenyl		55.5	0.58
Triphenylene		198.1	0.0
1,3,5-Tri-α-naphthylbenzene		197.3	0.40
1-Phenyl-naphthalene		unknown	
2-Phenyl-naphthalene			0.038

FIG. 6.1 Prefreezing anomalies of branched polyphenyls. (From Ubbelohde, 1978.)

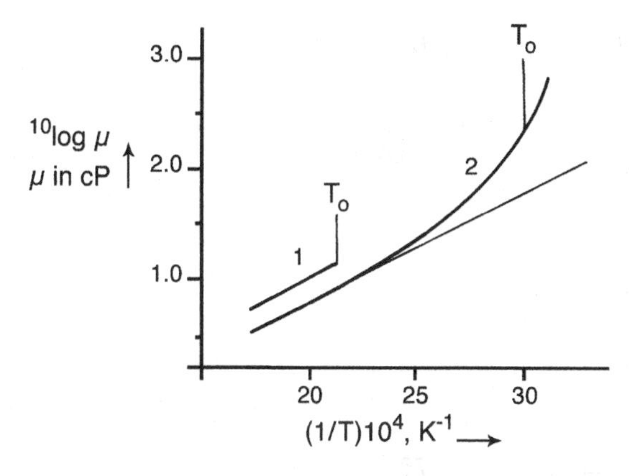

FIG. 6.2 Viscosity/temperature for triphenylene (1) and o-terphenyl (2). (From Ubbelohde, 1978.)

supercool. On approaching the melting point downward, the viscosity increases more than corresponds with the Andrade equation. Note that the viscosity at the melting point of the latter substance is approximately four times higher than the viscosity at the melting point of the two former compounds. The viscosity per se is a property promoting the occurrence of "anticrystalline" clusters. Also note that there is no alkyl group in the molecule of amino-diphenylamine, a small linear alkyl group in the molecule of 4-isopropyl-amino-diphenylamine, and a larger branched alkyl group in the molecule of the third substance. The behavior of 4-isopropyl-amino-diphenylamine is different from the behavior of amino-diphenylamine and the difference can be explained by the molecular configuration. The behavior of 4-isohexyl-amino-diphenylamine is different from the behavior of 4-isopropyl-amino-diphenylamine and this can be explained by the molecular configuration and the viscosity. Di-*p*-octylphenylamine will be discussed separately.

6.4.4. Viscosity Anomalies of Difunctional Alcohols

The substance with the simpler molecular configuration in Fig. 6.5 is much easier to crystallize than the substance with the slightly more complicated molecular shape. The latter compound shows a marked increase of the viscosity on approaching the melting point downward. Probably, the most important aspect is the viscosity per se. The simpler structure has a viscosity of $0.025\ \mathrm{N \cdot sec \cdot m^{-2}}$ at the melting point whereas the other compound has a viscosity of $0.6\ \mathrm{N \cdot sec \cdot m^{-2}}$ at the melting point. It is not easy for the molecules to align themselves to form a crystalline nucleus when the viscosity is that high. The two compounds are polyester additives.

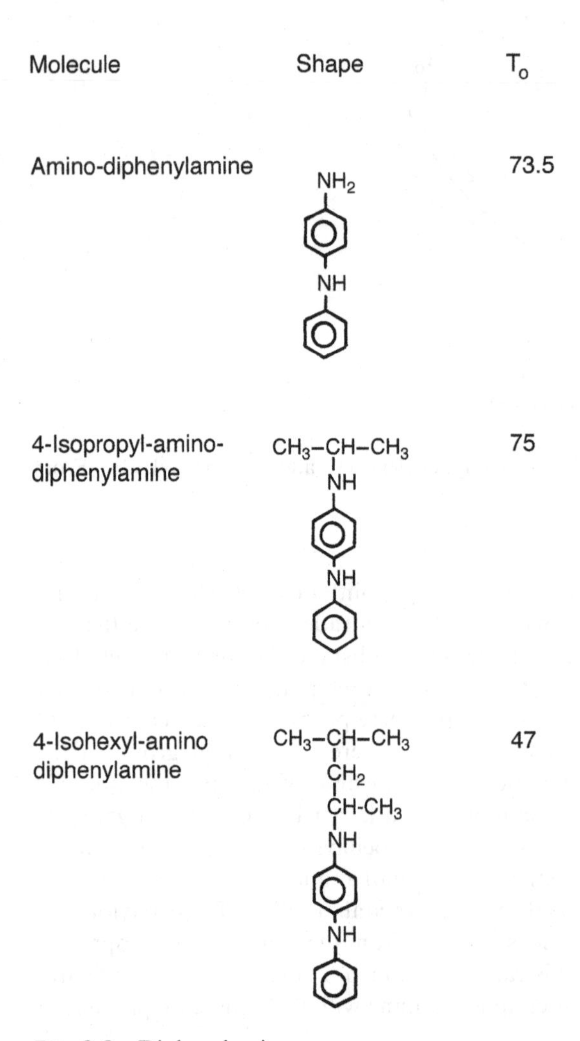

FIG. 6.3 Diphenylamines.

6.4.5. Viscosity Anomaly of Di-*p*-Octylphenylamine

Again, on approaching the melting point downward, the viscosity increases more than corresponds with the Andrade equation (Fig. 6.4). The molecular configuration of di-*p*-octyldiphenylamine is relatively complicated (Fig. 6.6). It was established that the melt behaves Newtonian, i.e., the viscosity depends on neither shear rate nor time. Fig. 6.7 is a plot of $^{10}\log \varphi$ as a function of

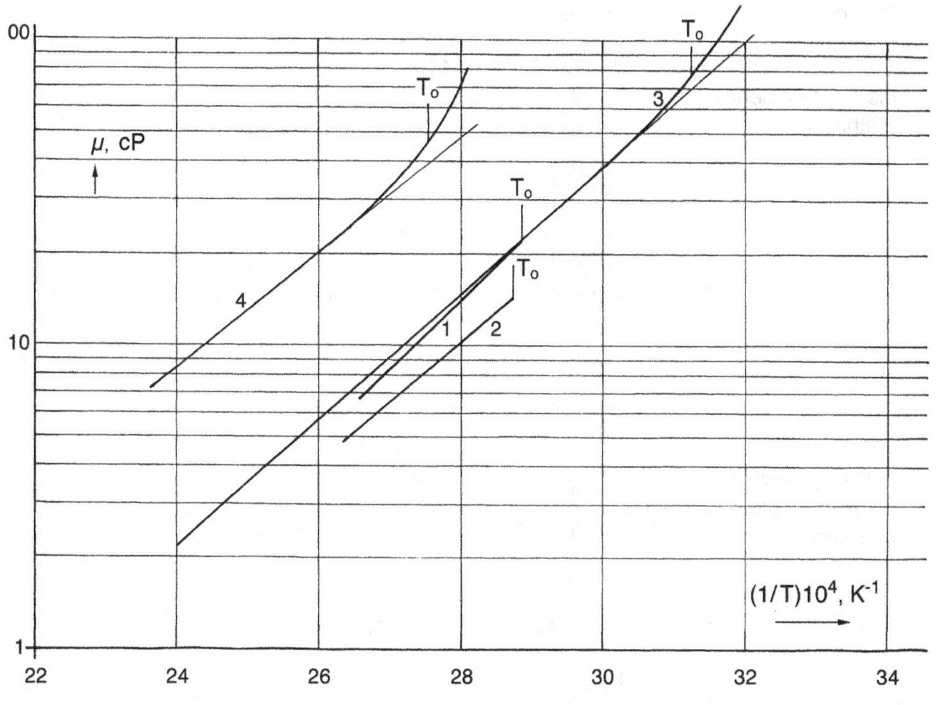

FIG. 6.4 Viscosity/temperature for amino-diphenylamine (1), 4-isopropyl-amino-diphenylamine (2), 4-isohexyl-amino-diphenylamine (3), and di-*p*-octylphenylamine (4).

$(1/T)10^4$. A straight line is obtained. Table 6.1 contains the values of μ_{act}, μ_{pred}, and φ. To arrive at the values, Eq. (6.2) was used.

6.5. EQUIPMENT

6.5.1. Sandvik Equipment

Fig. 6.8 is a process flow diagram of a Sandvik unit. At the left, a melt hold vessel is shown. Usually, the melt is kept under a nitrogen blanket to prevent discoloration. A centrifugal pump feeds the precrystallizer loop. Before entering this loop, the melt temperature is exactly adjusted to the temperature of the slurry in the loop to avoid upsetting the precrystallization system. The unit is started up by recycling the melt in the precrystallization system, material does not yet flow to the belt. The agitator of the precrystallizer rotates and cooling water is admitted to the jacket. Gradually, the melt in

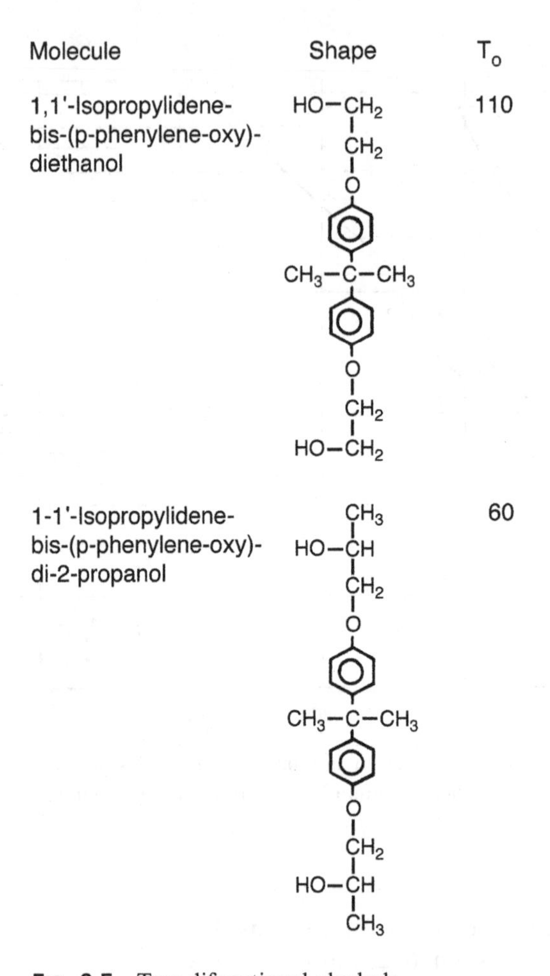

FIG. 6.5 Two difunctional alcohols.

the loop cools down to below the melting point. Under the combined action of cooling and agitation (shear), crystallization typically starts at a certain degree of supercooling and the temperature then rises rapidly to the melting point. The crystals and the liquid thus obtained have the same composition. The temperature rise is caused by the exothermicity of the crystallization process. Typically, for 4-isohexyl-amino-diphenylamine, the temperature rise is from 38 to 46°C. Continued cooling of the precrystallizer leads to an increase of the slurry density. This can be monitored by measuring, for example, the viscosity of the slurry. On attaining a certain viscosity, the slurry is passed on to the cooling belt via, for example, a Rotoform. The Rotoform is

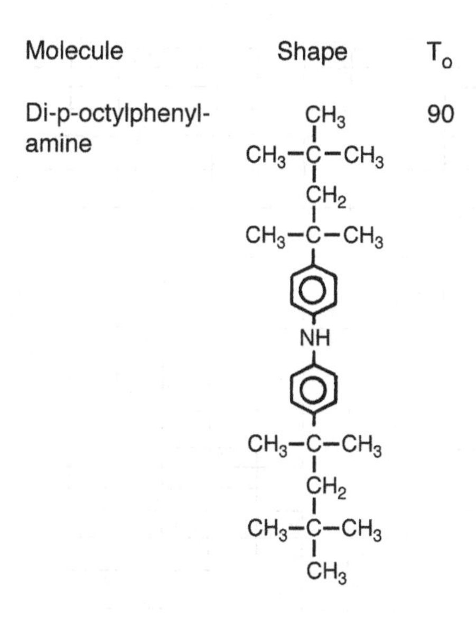

FIG. 6.6 Di-*p*-octylphenylamine.

a pastillating device described in Sec. 3.5. The precrystallized droplets solidify to pastilles. A stationary precrystallization process can be obtained by adjusting the melt flow to the loop and the cooling water temperature. The slurry flow in $kg \cdot hr^{-1}$ equals the melt flow in $kg \cdot hr^{-1}$.

It is even possible to control the slurry density automatically by adjusting the cooling water temperature. The agitator of the precrystallizer does not touch the wall. The clearance agitator/wall is approximately 1 mm. The precrystallizer loop, the transfer pump, the lines, and the Rotoform must all be traced carefully to avoid incrustation. Precrystallization is a stable process due to the heat of fusion. For example, 96.4 kJ must be carried away for the crystallization of 1 kg of 4-isohexyl-amino-diphenylamine. This heat must pass to the jacket and that requires time. Thus rapid changes of the slurry density cannot occur.

It is possible to adjust the belt cooling water temperatures. Substances having a strong tendency to supercool should be cooled initially with cooling water at, for example, 10 K below the melting point. The water in the other zones can be colder.

6.5.2. Twin-Screw Extruder

The discussion will be started by describing a single-screw extruder. A single-screw extruder can be compared to a nut and bolt. Let us assume that the bolt

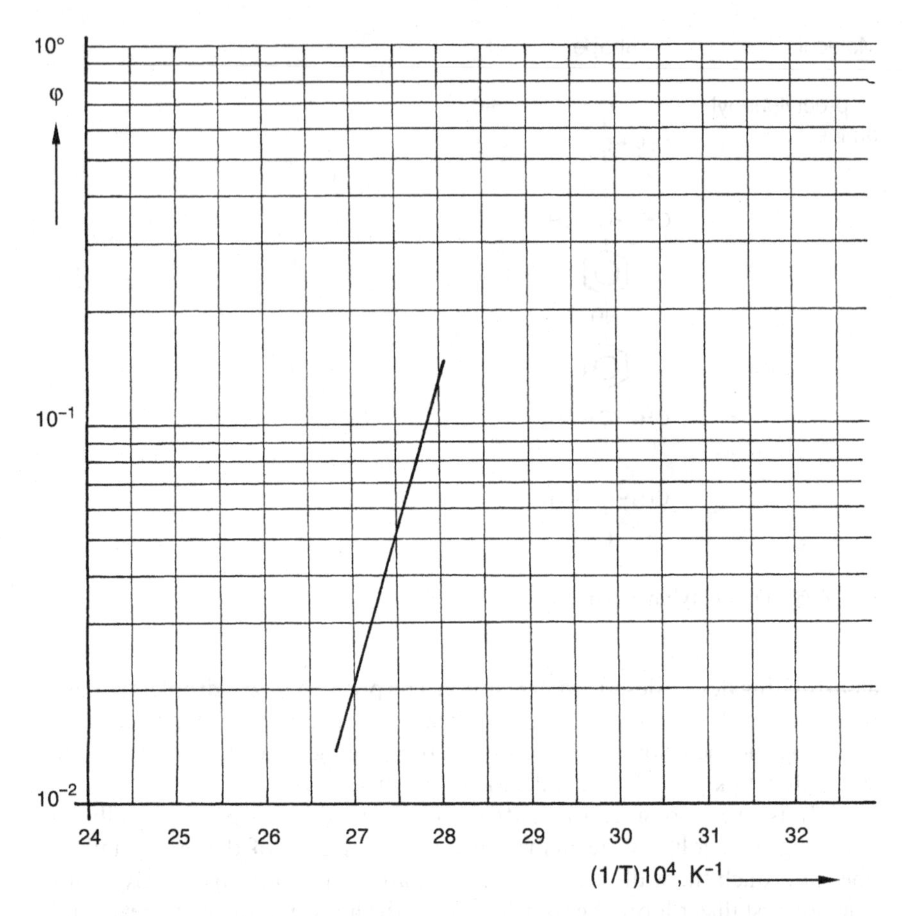

FIG. 6.7 Di-*p*-octylphenylamine: volume fraction φ blocked for flow by clustering.

TABLE 6.1 Di-*p*-Octylphenylamine. Ratio of the Volume of the "Anticrystalline" Clusters to the Total Volume (φ)

Temperature (°C)	μ_{act} ($N \cdot sec \cdot m^{-2} \cdot 10^{-3}$)	μ_{pred} ($N \cdot sec \cdot m^{-2} \cdot 10^{-3}$)	φ
105	24	24	0
100	29	28	0.015
95	36	33.5	0.0275
90	47	39	0.0625
85	65	46	0.1225

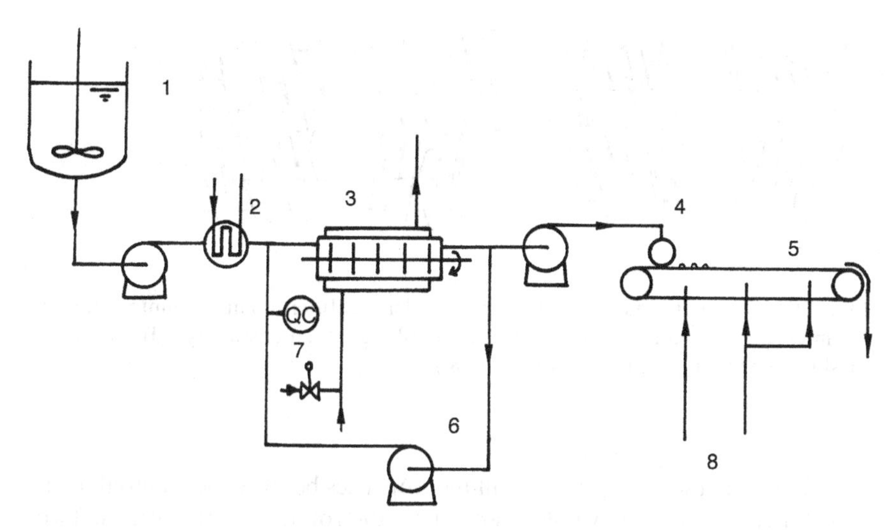

FIG. 6.8 Process flow diagram of a Sandvik unit for the pastillation of melts that tend to supercool. 1, melt hold vessel; 2, heat exchanger; 3, precrystallizer; 4, Rotoform; 5, cooling belt; 6, circulation pump; 7, viscosity measurement; 8, cooling water.

has a right-hand thread and one thread start only. The nut is on the bolt halfway the length. The nut cannot rotate, however, it can move up and down the bolt. On turning the bolt anticlockwise, the nut moves away from the head of the bolt. Similarly, a single-screw extruder having a right-hand thread and rotating anticlockwise transports a fluid from feed point to outlet. The transport is successful if the material processed slips at the screw surface and slides at the barrel surface. Coming back to the analogy with the nut and bolt, it is observed that the nut stays at the same position on the bolt if the nut rotates with the bolt. Similarly, there will be no output from a single-screw extruder if the process material sticks to the screw and slips at the barrel surface. Single-screw extrusion is not positive displacement and this sets severe limitations on single-screw extrusion.

In a twin-screw extruder two parallel screws are placed in a figure-of-eight section barrel. Generally speaking, twin screws can be divided into the two major categories of intermeshing and nonintermeshing screws. For nonintermeshing extruders, the separation between the screw axes is at least equal to the screw outer diameter. This configuration can be regarded in a way as two single-screw extruders which influence each other. Two categories of intermeshing screws can be distinguished: corotating screws and counter-rotating screws (Fig. 6.9). For counterrotating screws, the rotation can be such that they diverge above and converge below the intersecting area. This

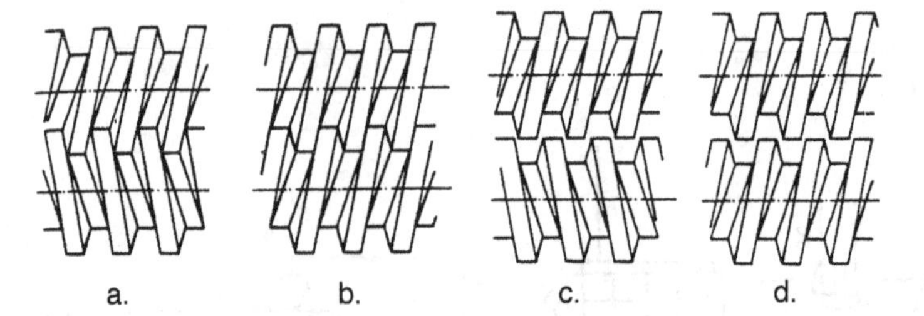

FIG. 6.9 Different kinds of twin screw extruders. a) Intermeshing, counterrotating; b) intermeshing, corotating; c) nonintermeshing, counterrotating; d) nonintermeshing, corotating. (From Janssen, 1978.)

choice confers the extruder good filling properties because the material fed is distributed through the whole chamber by the rotation of the screws. Two counterrotating intermeshing screws can be compared to two bolts. Looking in a direction from feed port to die, the right screw has, for example, a left-hand thread and a clockwise rotation whereas the left screw then has a right-hand thread and rotates anticlockwise. By the same token, two corotating screws rotate either clockwise both with the left-hand thread or anticlockwise both with the right-hand thread.

Intermeshing twin-screw extrusion is positive displacement. The material is conveyed by the leading edge of the screw in C-shaped chambers. In other words, intermeshing twin-screw extruders are self-cleaning (Fig. 6.10). Twin-screw extruders can have single-thread screws. It is also possible to work with screws having a number of thread starts. The output of twin-screw extruders is proportional to the number of thread starts, the rotational speed, and the chamber volume.

For mixing purposes, intermeshing corotating screws are preferred over intermeshing counterrotating screws. The reason is that corotating screws rotate in opposite directions in the contact area. This leads to high shear. Intermeshing counterrotating screws can be used in pumps. For the crystallization of melts that tend to supercool, the shear in intermeshing corotating screws is a bonus. Besides thread, it is possible to equip the shafts of twin-screw extruders with kneading sections. The kneading sections are lengths of corotating paddles of elliptical section, operating as a twin-screw extruder with a small positive displacement forward transport (Fig. 6.11).

It is even possible to work with sections that convey the material in the opposite direction. These sections cause a hold-up in the rate at which material moves along the screw. Often, the mixing and conveying systems can be supplied with "slip-on" sections and agitators, which readily permit the basic screw configuration to be altered.

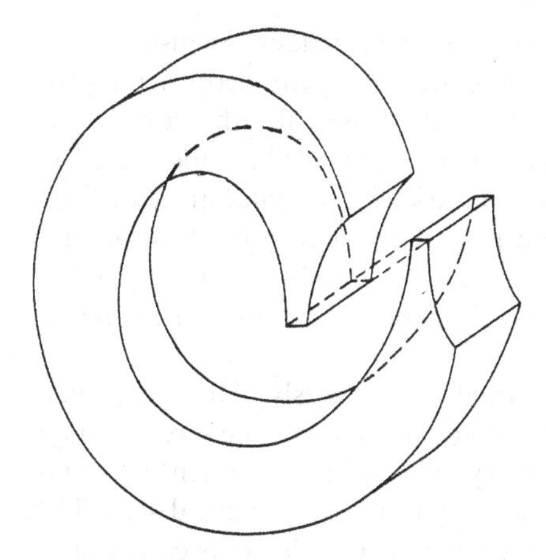

FIG. 6.10 A C-shaped chamber. (From Janssen, 1978.)

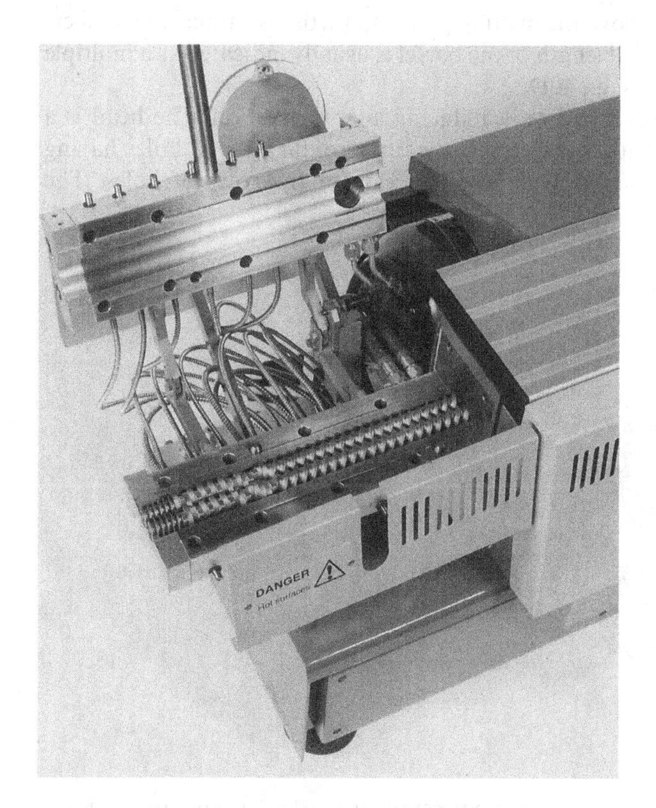

FIG. 6.11 Twin-screw extruder type MP19PC 15:1 with barrel open; the kneading section is visible. (Courtesy of APV Baker, Newcastle-under-Lyme, England.)

Fig. 6.12 depicts a large industrial twin-screw extruder. Twin-screw extruders with corotating intermeshing screws are widely applied for the processing of plastics. Furthermore, they are widely used for the continuous mixing of medium to high viscosity chemical and industrial compounds. Examples are adhesives, sealants, pet food, and doughlike materials. For food applications, the machines must have a sanitary and easy-to-clean design. Using a twin-screw extruder with corotating intermeshing screws for the crystallization of melts that tend to supercool is the application considered in this section.

The speed of the screws can be varied continuously. The screws can often be cooled, the area is comparable to the barrel area. The screws rotate while the cooling water lines are stationary. The sealing between lines and screws is by means of, for example, a stuffing box or mechanical seal. The barrel section is made up of a number of elements. The elements can be supplied individually with a cooling medium or a heating medium. On processing a melt that tends to supercool, the feed section of the barrel is held at a temperature slightly above the melting point. In further sections, the temperature is lowered. The full length of the barrel is usually indicated as a multiple of the screw diameter, e.g., 40D.

The die head is carried on a reducing/adaptor section. The head is a stainless steel block with, for example, two horizontal rows of holes having hot oil circulation ports drilled on either side of the rows of holes. The

FIG. 6.12 Twin-screw extruder type MP65MB. (Courtesy of APV Baker, Newcastle-under-Lyme, England.)

temperature of the oil is kept at a temperature slightly above the melting point. Thus the crystallized mass slides through the holes in a thin layer of liquid as the hot wall causes a small fraction of the material to melt. The feed section of the extruder is flooded. The product crystallizes and is extruded as rods. The rods are transferred onto a belt conveyor. The speed of the belt conveyor is slightly greater than the linear velocity of the rods leaving the die head. The rods are cooled by blowing air onto them. At the end of the belt, the rods are broken. On entering the storage, the material should be in the crystalline form for, for example, at least 99%. Material not fully crystallized may cake in the hopper. The production rate is determined by the heat transfer capability. Mainly, the heat of crystallization must be carried away.

At a fixed rotational speed, the torque of the machine can be used to control the feed rate. See the two examples. The combination of shear and low temperature produces seed crystals. The seed crystals subsequently grow while secondary nucleation takes place. Experience shows that a series of substances can thus be processed. The crystallization must be completed for, for example, 90% in a relatively short time. The installation of an industrial machine should be preceded by pilot plant tests. Scale-up is based on the heat exchanging area. The residence time in the industrial machine will rise because, on scale-up, $m^2 \cdot m^{-3}$ falls.

6.5.3. Drown-Out Equipment

Drown-out is the granulation of a melt in water. The granules produced have a porous structure and contain internal moisture which has to be removed. Experience has been obtained with batch processes. The independent process variables are:

- The amount of water
- The initial water temperature
- The final slurry density
- The stirrer type and the type of baffles
- The stirrer speed
- The melt feeding rate
- The presence of a surfactant.

Example

The Granulation of Di-*p*-Octylphenylamine (Melting Point 90°C)

One ton of di-*p*-octylphenylamine (Fig. 6.6) is granulated batchwise in an open cylindrical 9-m^3 vessel equipped with an anchor stirrer. The stirrer can rotate at two different speeds: 30 and 60 min^{-1}. The vessel is filled with 6.5 m^3 of water which is subsequently heated by means of live steam to 42°C. After the heating, the vessel contains 6.8 m^3 of water. The anchor stirrer speed is adjusted to 60 min^{-1}. One-and-a-half kilograms of dispersant is added to the drown-out pan. The melt is then run in over 20 min. The addition is stopped after 2 min and a sample of the batch is checked for satisfactory granulation. If the granulation is satisfactory the addition is restarted and completed. The temperature of the batch rises to 44–48°C. The poststir at 60 min^{-1} lasts 5–10 min. This poststir is followed by a second poststir at 30 min^{-1} lasting 1 hr. If the granulation is unsatisfactory the addition is restarted and live steam is passed into the batch to raise the temperature to 70°C. After completion of addition and settling, the lower aqueous layer is run to drain and the molten batch is run off to drums for blending away in subsequent manufacture.

After filtration and washing, the granules are dried at 50–55°C in a stove for 24 hr. The batch is discharged via a 0.5-in. stainless steel screen into paper bags.

6.6. EXAMPLE OF PASTILLE PRODUCTION

It is requested to scale-up pilot plant pastillation trials regarding 4-isohexyl-amino-diphenylamine. At the trials, 208 $kg \cdot hr^{-1}$ have been pastillated successfully. In the plant, an output of 600 $kg \cdot hr^{-1}$ is required.

6.6.1. Description of the Trials

Reference is made to Fig. 6.8. The melt hold vessel contains the molten material. The unit is started up by filling the precrystallizer including the recycle loop with molten material. Material does not yet flow to the belt. The melt flows through the recycle loop and the agitator of the precrystallizer is running. The lines and the pump are traced. Next, cooling water is admitted to the jacket and the contents of the system are cooled. The compound's melting point is 48°C. Crystallization starts when the process temperature is 39°C. The temperature rises to 48°C in 5 min due to the exothermicity of the crystallization. Now, slurry can be passed on to the Rotoform while melt is supplied from the melt vessel. A residence time of approximately 1 min on the belt is adjusted. Initially, the pastille shape is not fully satisfactory, i.e., the diameter is 6 mm while the height is 2.6 mm. On lowering the temperature of the cooling water to the jacket of the precrystallizer, the slurry density rises and the diameter of the pastille becomes 5.5 mm while the height rises to 2.75 mm. Stationary operation is now possible for up to 2 hr. The pastillated product shows an exotherm of 8 K in a Dewar in 10 min. This points to post-crystallization.

Physical Properties

$T_o = 48\,°C$
$i = 95{,}300\ J \cdot kg^{-1}$
$c_l = 2000\ J \cdot kg^{-1} \cdot K^{-1}$
c_s is assumed to be equal to c_l

6.6.2. Equipment Data

Precrystallizer

$L = 1300$ mm
$D = 250$ mm
$A = 1\ m^2$
$n = 66\ min^{-1}$

Rotoform

$L = 480$ mm
$D = 80$ mm
$d = 2$ mm
32 rows
54 holes per row
Hole pattern: see Fig. 6.13.

Cooling Belt

$L = 8.0$ m effective
$B = 480$ mm effective
$A = 3.84\ m^2$

6.6.3. Test Run Data

Temperatures, °C

Ambient	26
Melt hold vessel	55
Contents precrystallizer	48
Cooling water precrystallizer	10
Tracing	50
Cooling water belt zone I	off
Cooling water belt zone II	15
Cooling water belt zone III	15

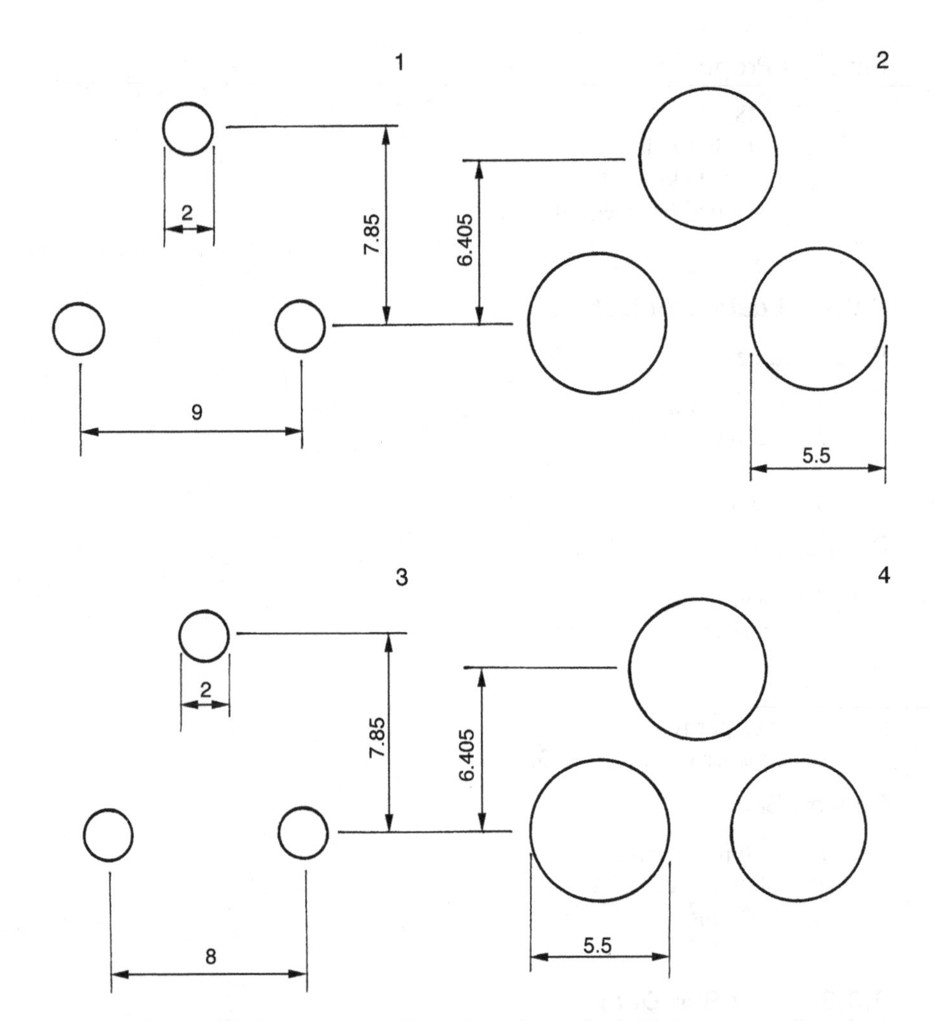

Fig. 6.13 Pastillation patterns (dimensions in mm). 1) Hole pattern pilot plant Rotoform; 2) pastille pattern pilot plant cooling belt; 3) hole pattern commercial Rotoform; 4) pastille pattern commercial cooling belt.

Product Data

Production	208 kg · hr^{-1}
Pastille diameter	5.5 mm
Pastille height	2.75 mm
Pastille weight	0.0452 g
Product exotherm	8 K (22°C to 30°C)

Miscellaneous Data

Motor agitator precrystallizer	4.4 A
Belt velocity	$9.1\ \text{m} \cdot \text{min}^{-1}$
Test time	105 min

6.6.4. Interpretation of the Test Run Data

Pastille Pattern on the Belt

Mass of one row of pastilles : $54 \cdot 0.0452 \cdot 10^{-3} = 0.00244$ kg

No. of rows on the belt per hr : $\dfrac{208}{0.00244} = 85,245.9$

Distance between rows : $\dfrac{60 \cdot 9.1}{85{,}245.9 - 1} = 0.006405$ m

(6.405 mm)

$$v_r = \frac{7.85}{6.405} \cdot 9.1 = 1.23 \cdot 9.1 = 11.2\ \text{m} \cdot \text{min}^{-1}$$

The circumferential velocity of the Rotoform is 1.23 times larger than the linear belt velocity (Fig. 6.13).

Residence Time on the Belt

$$\tau = \frac{8}{9.1} \cdot 60 = 52.7\ \text{sec}$$

6.6.5. Scale-Up

Precrystallizer

$600/208 \cdot 1 = 2.88\ \text{m}^2$, take a 5-$\text{m}^2$ precrystallizer (standard size).

Rotoform and Belt

L = 1080 mm (Rotoform)
D = 80 mm
d = 2 mm
32 rows
136 holes per row

Hole pattern: see Fig. 6.13.

$$v_r = 11.2 \cdot \frac{600}{208} \cdot \frac{54}{136} = 12.8 \text{ m} \cdot \text{min}^{-1}$$

$$v_b = \frac{12.8}{11.2} \cdot 9.1 = 10.4 \text{ m} \cdot \text{min}^{-1}$$

$$L = \frac{52.7}{60} \cdot 10.4 = 9.1 \text{ m, take 12 m (cooling belt).}$$

Pastille pattern: see Fig. 6.13

Miscellaneous Calculations Concerning the Pilot Plant

Initial slurry density in the precrystallizer

Crystallization starts when the process temperature is 39 °C. Because of the exothermicity of the crystallization process, the temperature rises to 48 °C. The temperature stabilizes at 48 °C because this is the melting point. It is assumed that the heating of the slurry is an adiabatic process. Actually, the heating lasts 5 min and cooling occurs. It is also assumed that the specific heat of the solid is equal to the specific heat of the liquid. The following equation is applicable:

$$2000(48 - 39) = \frac{a}{100} \cdot 95{,}300 \rightarrow a = 18.9\%$$

It could be that the slurry density is approximately 25% by weight.

Overall heat transfer coefficient in the precrystallizer

The heat flows are in $J \cdot hr^{-1}$.

Feed cooling 208 · 2000 (55–48)	= 2,912,000
Crystallization 208 · 0.25 · 95,300	= 4,955,600 +
	7,867,600
Heat input by the tracing 0.1 · 7,867,600	= 786,760 +
	8,654,360
Heat input by the agitator $0.5\,EI\sqrt{3}\cos\varphi \cdot 3600 = 0.5 \cdot 380 \cdot 4.4 \cdot \sqrt{3} \cdot 0.8 \cdot 3600$	= 4,170,224 +
	12,824,584

$$U = \frac{12{,}824{,}584}{1(48 - 10)3600} = 93.7 \text{ W} \cdot \text{m}^{-2} \cdot \text{K}^{-1}$$

It is assumed that 50% of the agitator power consumption is dissipated in the process stream.

The heat transfer coefficient of a 1-mm layer in the clearance between agitator and wall is approximately $\lambda_s/d = 0.15/10^{-3} = 150\ \mathrm{W \cdot m^{-2} \cdot K^{-1}}$.

It could be that the main resistance to heat transfer is located in the solid layer in the clearance. It is hence important to keep the clearance constant on scaling-up.

Belt utilization

See Fig. 6.13.
Area triangle $0.5 \cdot 9 \cdot 6.405 = 28.82\ \mathrm{mm^2}$
Area pastilles in triangle $0.5(\pi/4)5.5^2 = 11.88\ \mathrm{mm^2}$
$(11.88/28.82)100 = 41.2\%$
kg per $\mathrm{m^2}$ belt area: $0.5 \cdot 0.0452 \cdot 10^{-3}/(28.82 \cdot 10^{-6}) = 0.784$
Production intensity: $208/3.84 = 54.2\ \mathrm{kg \cdot m^{-2} \cdot hr^{-1}}$

Degree of crystallization of the product

The product exhibits an exotherm of 8 K.
Percentage liquid: $(1 \cdot 2000 \cdot 8/95{,}300)100 = 16.8$

The product is not very hard on leaving the belt. It is good practice to condition the product on a belt conveyor. The layer thickness could be 10 cm to avoid caking while the residence time is 30 min.

Slurry density measurement by means of calorimetry (example)

A calorimeter is filled with 1000 g of water.
Calorimeter and water are at 20.0°C.
The water value of the calorimeter is $100\ \mathrm{J \cdot K^{-1}}$.
A total of 546 g of slurry from the precrystallizer is added.
The final temperature is 33.1°C.
Heat absorbed = heat released

$$(1000 \cdot 4.2 + 100)(33.1 - 20.0) = 0.546\{2000(48 - 33.1) + 95{,}300(1 - a/100)\}$$

$a = 23.0 \rightarrow$ the slurry density is 23.0% by weight.

6.7. EXAMPLE OF ROD PRODUCTION (1)

4-Isohexyl-amino-diphenylamine is processed successfully into rods by means of a twin-screw extruder. A series of experiments is carried out to establish the maximum capacity.

6.7.1. Description of the Trials

The melt is pumped from the melt hold vessel to the extruder. The liquid is pumped through the extruder by the rotating screws. Both the screws and the jacket are cooled. At some point along the extruder, crystallization starts under the combined action of cooling and shear. The slurry is transported further and the slurry density gradually increases. Finally, the product is extruded through 2-mm holes. The degree of crystallization is then in the range 90–100%. The extruded product (the rods) falls on a conveyor belt and is cooled by means of air. The rods are processed through a cutter and stored. The capacity of the extruder is measured by varying both the feed and the rotational speed.

Physical Properties

T_o = 48°C
i = 95,300 $J \cdot kg^{-1}$
c_l = 2000 $J \cdot kg^{-1} \cdot K^{-1}$
c_s is assumed to be equal to c_l

6.7.2. Equipment Data

Make:	Werner & Pfleiderer
Type:	C120, twin-screw intermeshing corotating
Screw diameter:	120 mm
Screw length:	40D, i.e., 4.80 m
Cooling:	both screws and jacket cocurrent flow
Cooled area screws:	$0.784\ m^2 \cdot m^{-1}$ $3.7632\ m^2$
Cooled area jacket:	$0.603\ m^2 \cdot m^{-1}$ $2.8944\ m^2$
Rotational speed:	10.4–104 min^{-1}
Motor:	42 kW
Power consumption:	42 kW at 104 min^{-1} and 100% torque Proportional to both rotational speed and torque
Number of holes:	140 (2 rows, 70 holes per row)
Hole diameter:	2 mm

6.7.3. Test Run Data

Temperatures, °C

Melt hold vessel	65
Rods	47
Cooling water inlet	8
Cooling water outlet	16

Production

No.	min^{-1}	$kg \cdot hr^{-1}$	% torque	extrusion
1	15	330	36	good
2	22	456	41	good
3	25	468	41	good

See Figs. 6.14 and 6.15.

Miscellaneous data
Assumptions

Power consumption empty machine:	0.5 kW
Degree of product crystallization:	90%

6.7.4. Interpretation of Test Data

Raising the speed of the screws from 15 to 22 min^{-1} leads to a pro-rata increase of the production from 330 to 456 $kg \cdot hr^{-1}$. The torque also rises. However, raising the rotational speed from 22 to 25 min^{-1} hardly results in a capacity increase. We are approaching the limits of the machine as it is not possible to carry away more heat to crystallize more material. On raising the speed beyond 25 min^{-1} and the feed beyond 470 $kg \cdot hr^{-1}$ the operation will collapse and "wet" material will be extruded. The rods will stick together and the cutter will break down.

6.7.5. Calculation of U at 22 min^{-1}

Heat transferred, $kJ \cdot hr^{-1}$

Feed cooling 456 · 2.0 (65–48)	= 15,504
Crystallization 456 · 0.9 · 95.3	= 39,111
Product cooling 456 · 2.0 · 1	= 912
Power input by screws {42(41/100)(22/104) – 0.5}3600	= 11,314 +
	66,841

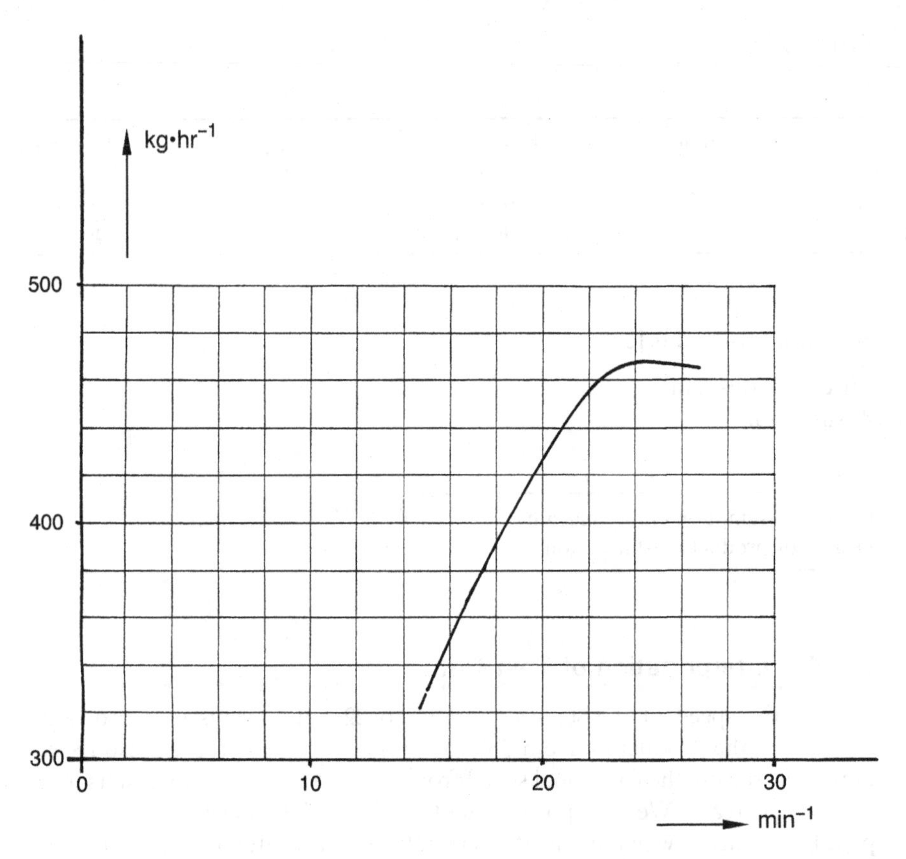

FIG. 6.14 $kg \cdot hr^{-1}/min^{-1}$ for rod production (1).

$$\Delta T_m = \frac{(65-8)-(47-16)}{{}^{e}\log\frac{65-8}{47-16}} = 42.7\ K$$

$$A = 6.6576\ m^2$$

$$U = \frac{66,841,000}{3600 \cdot 42.7 \cdot 6.6576} = 65.3\ W \cdot m^{-2} \cdot K^{-1}$$

Work input

$$\frac{11,314,000/3600}{456} = 6.89\ Whr \cdot kg^{-1}$$

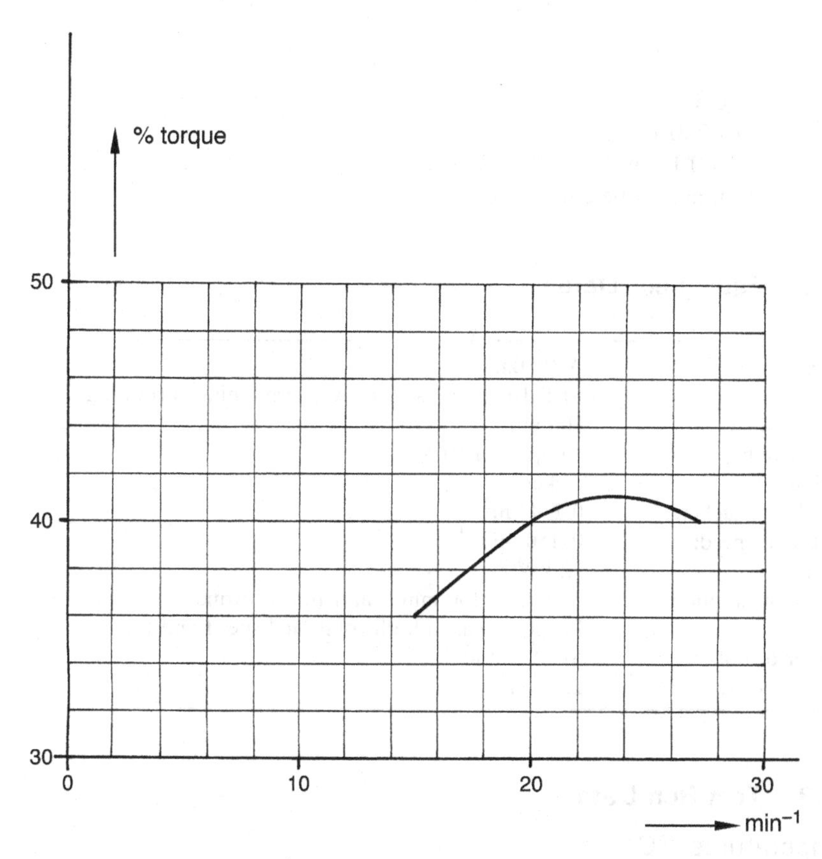

Fig. 6.15 Torque/min^{-1} for rod production (1).

6.8. EXAMPLE OF ROD PRODUCTION (2)

Di-*p*-octylphenylamine is processed successfully into rods by means of a twin-screw extruder. The capacity is 100 kg · hr^{-1} and it is desired to raise the capacity.

6.8.1. Description of the Trials

See the first example of the production of rods by means of an extruder. Here, the jacket is cooled only.

Physical Properties

$T_o = 90°C$
$i = 61{,}750\ J \cdot kg^{-1}$
$c_l = 2240\ J \cdot kg^{-1} \cdot K^{-1}$
c_s is assumed to be equal to c_l

6.8.2. Equipment Data

Make:	APV Baker
Type:	M-P 100 mm, twin-screw intermeshing corotating
Screw diameter:	100 mm
Screw length:	15D, i.e., 1.50 m
Cooling:	jacket
Cooled area jacket:	0.7297 m^2
Rotational speed:	0–150 min^{-1}
Motor:	22 kW
Power consumption:	22 kW at 150 min^{-1} and 100% torque Proportional to both rotational speed and torque
Number of holes:	51 (2 strands)
Hole diameter:	2.38 mm

6.8.3. Test Run Data

Temperatures, °C

Melt hold vessel	106
Rods	75
Cooling water	16

Production

No.	min^{-1}	$kg \cdot hr^{-1}$	% torque	extrusion
1	10	62	50	good
2	12	84	45	good
3	15	106	35	reasonable
4	20	128	23	bad

See Figs. 6.16 and 6.17.

Miscellaneous data
Assumptions

Power consumption empty machine:	0.5 kW
Degree of product crystallization:	90%

6.8.4. Interpretation of Test Data

Raising the speed of the screws from 10 to 15 min^{-1} leads to a pro-rata increase of the production from 62 to 106 $kg \cdot hr^{-1}$. The torque falls. However, raising the speed to 20 min^{-1} leads to only a small capacity increase. We are approaching the limits of the machine. Further raises in speed and throughput will lead to "wet" extrusion.

6.8.5. Calculation of U at 12 min^{-1}

Heat transferred, $kJ \cdot hr^{-1}$

Feed cooling 84 · 2.24 (106–90)	= 3011
Crystallization 84 · 0.9 · 61.75	= 4668
Product cooling 84 · 2.24 (90–75)	= 2822
Power input by screws	
{22(45/100)(12/150) – 0.5}3600	= 1051 +
	11,552

$$\Delta T_{\mathrm{m}} = \frac{(106 - 16) - (75 - 16)}{{}^{e}\log \dfrac{106 - 16}{75 - 16}} = 73.4 \text{ K}$$

$$A = 0.7297 \text{ m}^2$$

$$U = \frac{11,552,000}{3600 \cdot 73.4 \cdot 0.7297} = 59.9 \text{ W} \cdot \text{m}^{-2} \cdot \text{K}^{-1}$$

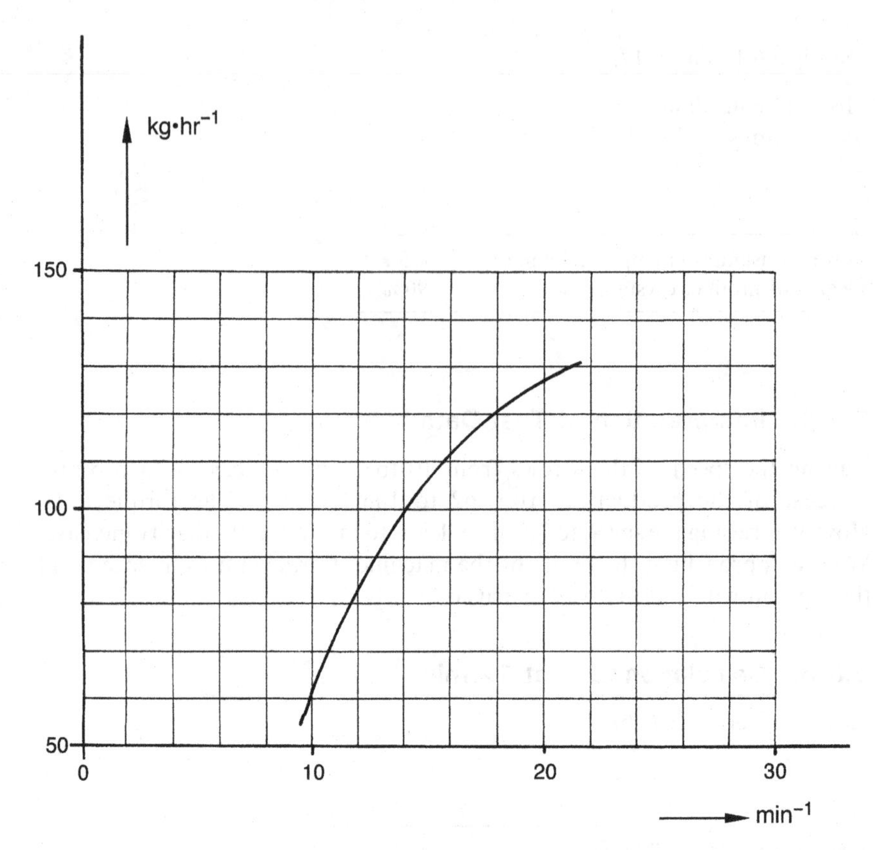

FIG. 6.16 $kg \cdot hr^{-1}/min^{-1}$ for rod production (2).

Work input

$$\frac{1{,}051{,}000/3600}{84} = 3.48 \text{ Whr} \cdot kg^{-1}$$

6.8.6. Capacity Extension

On cooling the jacket with a cooling medium at –20°C, the capacity can be extended to approximately 120 $kg \cdot hr^{-1}$. The feasibility of using a temperature of –20°C should be checked with the equipment manufacturer. The cooling of the screws is a further possibility. Finally, the installation of a scraped-surface heat exchanger between the melt hold vessel and the extruder can be considered. In the scraped-surface heat exchanger, a precrystallization can be accomplished.

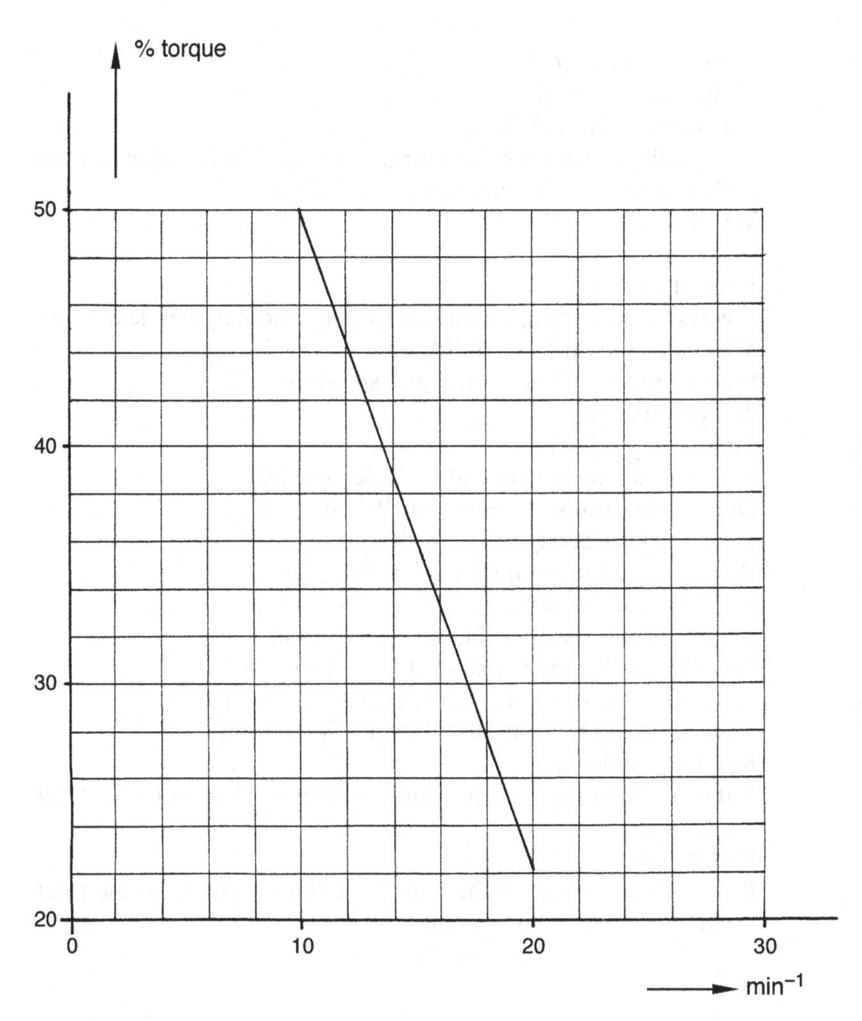

FIG. 6.17 Torque/min^{-1} for rod production (2).

LIST OF SYMBOLS

A	Constant in Eq. (1.1)
	Precrystallizer area, cooling belt area or extruder area [m^2]
A'	Constant in Eq. (6.1)
a	Crystal slurry density [% by weight]
B	Constant in Eq. (6.1)
	Cooling belt width [m]

B'	Constant in Eq. (6.2)
c_l	Melt specific heat $[J \cdot kg^{-1} \cdot K]$
c_s	Solid specific heat $[J \cdot kg^{-1} \cdot K]$
D	Precrystallizer diameter, Rotoform diameter or screw diameter [m]
d	Hole diameter or clearance agitator/wall [mm]
E	Tension [V]
I	Current [A]
i	Heat of fusion $[J \cdot kg^{-1}]$
L	Precrystallizer length, Rotoform length or cooling belt length [m]
M	Molecular weight $[g \cdot gmol^{-1}]$
n	Precrystallizer agitator rotational speed $[min^{-1}]$
T	Temperature [K]
T_o	Melting point [°C]
ΔT_m	Logarithmic mean temperature difference [K]
U	Overall heat transfer coefficient $[W \cdot m^{-2} \cdot K^{-1}]$
v_b	Cooling belt velocity $[m \cdot min^{-1}]$
v_r	Rotoform circumferential velocity $[m \cdot min^{-1}]$
λ_s	Solid thermal conductivity $[W \cdot m^{-1} \cdot K^{-1}]$
μ	Melt dynamic viscosity $[N \cdot sec \cdot m^{-2}$ or cP]
μ_{act}	Actual melt dynamic viscosity $[N \cdot sec \cdot m^{-2}]$
μ_{pred}	Predicted melt dynamic viscosity $[N \cdot sec \cdot m^{-2}]$
ρ_l	Liquid specific mass at 20°C $[g \cdot cm^{-3}]$
τ	Residence time [sec]
φ	Ratio of the volume of the "anticrystalline" clusters to the total volume
	Power angle
φ_0	Ratio of the volume of the "anticrystalline" clusters to the total volume at the melting point

REFERENCES

Janssen, L. P. B. M. (1978). *Twin Screw Extrusion*. Amsterdam, The Netherlands: Elsevier.

Poling, B. E., Prausnitz, J. M., O'Connell, J. P. (2000). *The Properties of Gases and Liquids*. New York: McGraw-Hill.

Ubbelohde, A. R. (1978). *The Molten State of Matter*. Chichester, England: Wiley.

7

Melt Crystallization

7.1. INTRODUCTION

Melt crystallization is a separation technique. It comprises the cooling crystallization of a material from a melt without using a solvent. It can be an alternative to distillation when the boiling points of the components in a melt are close together. Furthermore, melt crystallization has the potential of ultrapurification. For example, the feed contains 70–80% by weight of the desired material and the product has an assay of 99.99% by weight. To attain this purity, melt crystallization has to be repeated. The subject of melt crystallization is introduced in Sec. 7.2.

Phase diagrams are discussed in Sec. 7.3. The phase diagram is important with regard to the efficiency and applicability of melt crystallization. Melt crystallization can be applied when the phase diagram is of the eutectic type. Eutectic systems may be formed if the components are different in size and shape on a molecular scale. For example, *p*-xylene and *m*-xylene form a eutectic mixture. It is more difficult to apply melt crystallization when the phase diagram is of the solid solution type. Solid solutions may be formed if the components are similar in size and shape on a molecular scale.

Constitutional supercooling is discussed in Sec. 7.4. This phenomenon explains the existence of an inherent tendency for irregular growth in many instances. Irregular growth can cause occlusions and thus lower the purity of the solid material. Regular growth is promoted by good mixing and low growth rates. Layer melt crystallization and suspension melt crystallization are the two purification methods carried out commercially. Equipment for layer melt crystallization is relatively easy to design and control. Layer melt crystallization is the most important method. The technique is discussed in Sec. 7.5. A distinction can be made between static and dynamic methods. At the static method, the material is crystallized batchwise on a heat-exchanging surface from a stagnant melt in a closed vessel. Because of poor mixing, the growth rate has to be low to attain the desired purity. A low growth rate means a long residence time. Typically, the crystallization time is 12 hr. Vertical single-pass tube-and-shell heat exchangers are used for dynamic layer crystallization. The melt is pumped through the tubes and the crystalline layer grows on the inside of the tubes. The cooling medium is passed through the shell. Dynamic layer crystallization is also carried out batchwise. Because of good mixing, the growth rate can be higher than for static layer crystallization. Typically, a crystallization lasts an hour.

Finally, suspension melt crystallization is discussed in Sec. 7.6. *p*-Xylene and *p*-dichlorobenzene, for example, are commercially recovered continuously by means of this technique. Suspension melt crystallization is more difficult to design and control than layer melt crystallization. The reasons are the necessity to handle slurries, the need to control the particle size distribution, and the need to carry out liquid/solid separations. Because of these inherent aspects, suspension melt crystallization is only used for the purification of a limited number of organic chemicals. Suspension melt crystallization can also be carried out batchwise.

7.2. INTRODUCTORY REMARKS CONCERNING MELT CRYSTALLIZATION

If a drum filled with, for example, liquid caprolactam is stored in a warehouse at ambient temperature, crystallization will start at the vessel wall. A well-defined crystal–melt interface progresses through the melt and impurities are rejected into the liquid phase. By the same token, the caprolactam crystallizing first is purer than the material crystallizing later on. This principle can be applied industrially and the resulting technique is called normal or progressive freezing. Rittner and Steiner (1985) describe the purification of monochloroacetic acid by Hoechst by means of layer crystallization since 1901. This is the first known industrial application of melt crystallization. The purified material is used for the production of indigo dyestuffs. A vessel is filled with relatively impure monochloroacetic acid. Cooling water flows through a number of heat exchanger tubes present in the vessel. The purified material crystallizes on the outside of the tubes. The cooling is stopped and the remaining melt is withdrawn through the bottom outlet when a certain mass fraction has crystallized. Next, a sweating process is induced by a slight temperature increase caused by the flow of warmer water through the tubes. Thus, an additional upgrading is obtained. Finally, the purified material is molten. Equipment for layer crystallization does not contain moving parts and there is no need for solid/liquid separation devices. Long residence times are required because the mass transfer occurs by natural convection and diffusion, and high growth rates would result in impure materials. Long residence times lead to large equipment. The aforementioned crystallization process could be carried out in a stirred vessel as well. Crystals can grow in suspension. This crystallization process can take place faster than the layer crystallization process because the crystal area per cubic meter vessel volume is much larger. However, the suspended crystals must be separated from the mother liquor by means of liquid/solid separation equipment. Furthermore, moving parts (e.g., stirrers and pumps) are needed for this option. Melt crystallization is applied for the purification and ultrapurification of organic chemicals. It is an alternative for distillation and extraction—two important separation techniques.

On comparing distillation and melt crystallization, it can be stated that the heat of evaporation (in Joules per kilogram) for a given compound is greater than the heat of fusion. A factor of three is not uncommon. Furthermore, the process temperatures at distillation are generally much higher than those at melt crystallization. For melt crystallization, low-level energy usually suffices. However, some melt crystallizations take place at subzero temperatures (e.g., the separation of *p*-xylene from a mixture of ethylbenzene, *p*-xylene, *o*-xylene, and *m*-xylene occurs industrially at $-40/-65\,°C$). It is interesting to note that pure *p*-xylene melts at $+13\,°C$. A further aspect is that the melt crystallization process must often be repeated several times. Finally, the heat supplied in the reboiler of a distillation column can be recovered at a lower level in the reflux condenser. Because of all these aspects, it cannot be stated that the variable costs of melt crystallization are always lower than the variable costs of distillation. Each case must be considered per se.

The distribution coefficient, usually indicated by the symbol k, is the fundamental descriptive parameter of crystallization. It is the ratio of the solute concentration in the solid to the solute concentration in the melt from which the solid is formed. Conventional phase diagrams are graphical representations of the sequence of k-values measured for a series of liquids of increasing solute content. However, the ideal thermodynamic separation effect is impaired by occlusion of liquid during the crystallization and by the presence of residual melt adhering to the solid after solid/liquid separation.

As stated, the process temperatures at melt crystallization are relatively low. This is favorable for the stability of organic compounds.

Melt crystallization is very good with regard to process safety and ecology. Containment is easier than for, for example, distillation. Different from crystallization from a solvent, an auxiliary material (i.e., the solvent) is not needed.

A typical growth velocity at the melt crystallization of organic compounds is 10^{-7} m · sec^{-1}. Crystallizing metals have growth velocities up to 10^{-3} m · sec^{-1}, whereas ionic crystals grow at typically 10^{-5} m · sec^{-1}.

During melt crystallization, the execution of a countercurrent process is not straightforward as the diffusion coefficient in the solid phase is typically 10^{-12} m^2 · sec^{-1}. For the gas phase, 10^{-5} is a typical value, whereas 10^{-9} is applicable for liquids.

7.3. PHASE DIAGRAMS

Solid–liquid phase diagrams represent the relationship between temperature and composition. If a phase diagram is known, a good impression of the separability of the components of a mixture by a crystallization process can be obtained. The discussion will be restricted to phase diagrams obtained at atmospheric pressure. Furthermore, binary mixtures will be considered only as the diagrams become more complex when there are more than two components. Industrial samples, however, are usually multicomponent mixtures.

It is possible to classify the phase diagrams into three groups:

- The components are immiscible in the solid state—eutectic systems.
- The components are miscible in the solid state—solid solutions.
- There is partial miscibility in the solid state.

The two first groups will be discussed shortly.

7.3.1. Eutectic Systems

Components Immiscible in the Solid State

Matsuoka and Fukushima (1986) collected data of binary systems and concluded that more than 80% of the systems found were eutectic. Eutectic systems may be formed if the two components of a binary mixture are different in size and shape on a molecular scale. Fig. 7.1 shows the phase diagram of a binary system forming a simple eutectic. The melting points of the substances A and B are $T_{o,A}$ and $T_{o,B}$, respectively. The curve $T_{o,A}ET_{o,B}$ is called the liquidus. It indicates the lowest temperature at which the liquid of a given composition can exist as a single phase. The ideal crystallization behavior of a mixture having composition $X_{B,2}$ is depicted in Fig. 7.1. First, the mixture is cooled from T_C to T_D. Pure B crystals appear in the mixture at T_D. Further cooling results in more B crystals, whereas the composition of the liquid follows the liquidus. At T_E, the remaining liquid solidifies as two pure

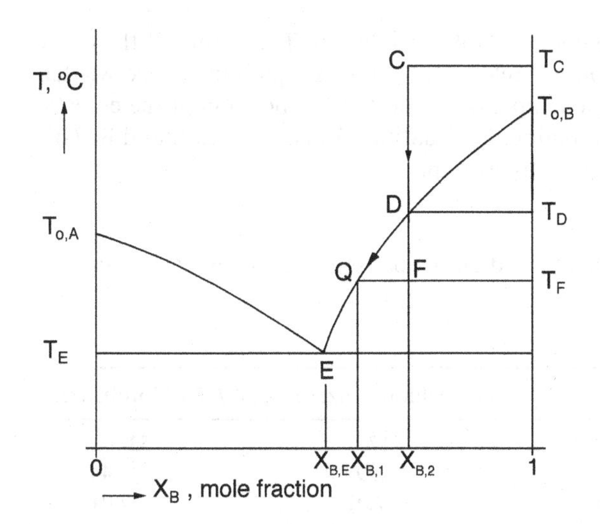

FIG. 7.1 Phase diagram of a binary system forming a simple eutectic.

solid phases A and B according to the eutectic composition $X_{B,E}$. Eutectic means "easily melted." The crystallization behavior of the eutectic composition resembles that of a pure compound, solidifying completely at T_E. The eutectic mixture contains very fine A crystals and very fine B crystals. In F, the ratio of the number of crystalline B moles to the number of moles in the liquid phase is QF/FT_F. This geometrical analogy is known as the lever rule. This will be proven.

In F, there are m_B moles of solid B and m_L moles of liquid. The total number of moles in the sample is $(m_B + m_L)$. The number of A moles in the sample is $(1 - X_{B,2})\ (m_B + m_L)$.

It is also $(1 - X_{B,1})m_L$.

$$(1 - X_{B,2})(m_B + m_L) = (1 - X_{B,1})m_L$$

It follows $m_B/m_L = (X_{B,2} - X_{B,1})/(1 - X_{B,2}) = QF/FT_F$.

The line QFT_F is called the tie-line.

The shape of the liquidus can be calculated if A and B form ideal solutions. Furthermore, the heats of fusion and the melting points of the two components must be known. The relevant equation is:

$$^{e}\log X_A = \frac{i_A}{R}\left(\frac{1}{T_{o,A}} - \frac{1}{T}\right) \tag{7.1}$$

This is the Van 't Hoff expression. It enables the calculation of the melting point depression substance A experiences in a solution. When an ideal solution containing a mole fraction X_A of substance A is cooled to temperature T, solid A starts to come out of the solution. The nature of the other component in the solution is irrelevant. A similar equation can be written down for substance B. The temperature and the composition of the eutectic point can be found by combining the equations for substances A and B. This will be illustrated by means of an example.

Example 1

o-Dichlorobenzene and *p*-dichlorobenzene form an ideal solution (Muir, 1982).

	o-Dichlorobenzene	*p*-Dichlorobenzene
T_o (°C)	−17	53
i ($J \cdot kg^{-1}$)	87,990	124,440
Molecular weight ($kg \cdot kmol^{-1}$)	147.01	147.01

At the eutectic point:

$$^{e}\log X_A = \frac{87{,}990 \cdot 147.01}{8314}\left(\frac{1}{256} - \frac{1}{T_E}\right)$$

$$^{e}\log(1 - X_A) = \frac{124{,}440 \cdot 147.01}{8314}\left(\frac{1}{326} - \frac{1}{T_E}\right)$$

It follows $X_A = 0.87$ and $T_E = -23°C$.
The actual values are $X_A = 0.86$ and $T_E = -20.5°C$.

A purification of compound B is only possible when the feed contains a greater concentration of B than the eutectic composition. A high purity of separated B in one single crystallization step is possible. However, industrially, the crystallization is not carried out under equilibrium conditions. Kinetic effects may affect the separation efficiency and it will be difficult to obtain 100% pure B.

Eutectic with Polymorphism of One Component

This phenomenon is illustrated in Fig. 7.2. If a liquid having composition C is cooled, pure crystalline B is obtained in form I. When the remaining liquid attains the composition and temperature defined by point P, all crystalline B undergoes isothermal transition to form II. This transition is associated with evolution of the latent heat of transition. Further cooling yields crystals of B in form II until the eutectic point is attained. Too rapid cooling at P may lead to supercooling of form I as indicated by the dashed line. The feasibility of

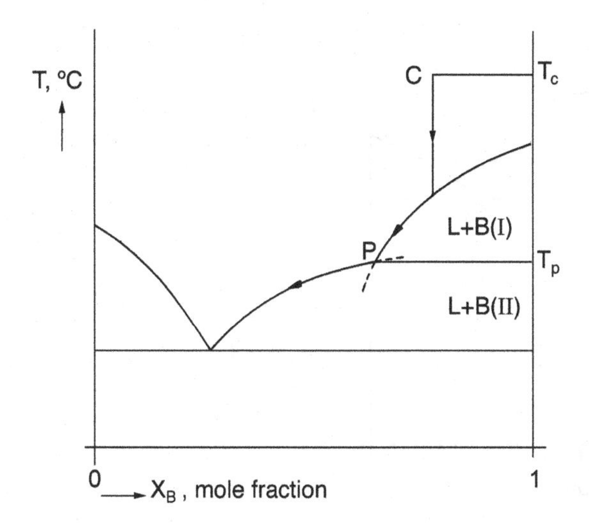

FIG. 7.2 Eutectic with polymorphism of one component.

purification by means of melt crystallization is comparable to the case of the simple eutectic.

Eutectic with a Congruently Melting Compound

This phenomenon is illustrated in Fig. 7.3. The two compounds A and B interact to form a new compound of composition A_XB_Y. X and Y are usually small integers. A_XB_Y has a melting point that may be higher or lower than the melting points of A and B. The system may be considered as the combination of two simple eutectic systems. The *m*-cresol/*p*-cresol system is a well-known example of the occurrence of such a molecular compound. Besides the three types described, there are two further main types.

7.3.2. Solid Solutions

Solid solutions may be formed if the two components of a binary mixture are similar in size and shape on a molecular scale. In the case of binary organic mixtures, host molecules in the crystal lattice are replaced by guest molecules. Host molecules may be replaced more easily by smaller guest molecules than by larger ones. For mutual solid solubility to be continuous throughout the full range of composition, three conditions must be satisfied:

- The structures of the components must have identical space groups.
- The two different unit cells concerning the crystallization of the two components per se must have the same number of molecules.
- The two molecular packings must be similar.

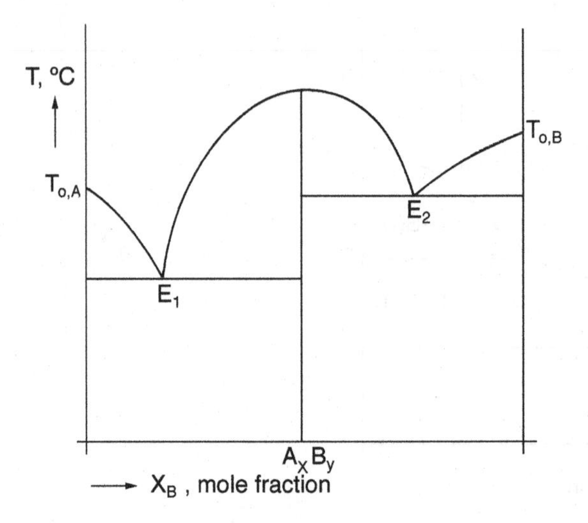

FIG. 7.3 Eutectic with a congruently melting compound.

However, a number of cases are known in which continuous solid solubility exists although the conditions are not satisfied. Fig. 7.4 shows continuous mutual solid solubility. The melting points of the pure compounds define the ends of the liquidus and solidus curves. On cooling a liquid having composition $X_{B,2}$, a solid having composition $X_{B,1}$ crystallizes at temperature T_D. On continued cooling, the composition of the liquid follows the liquidus,

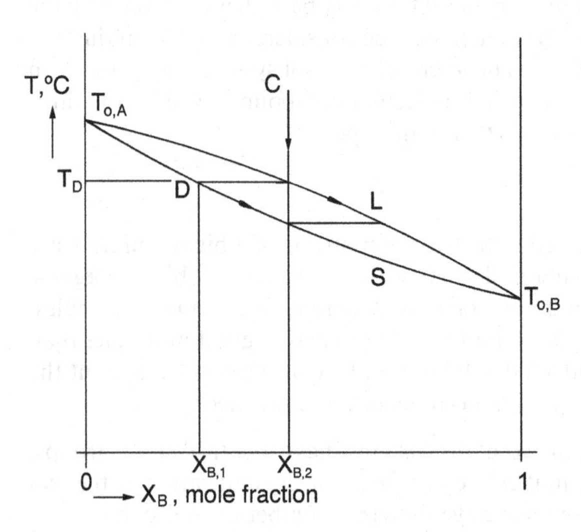

FIG. 7.4 Phase diagram of a binary system forming solid solutions.

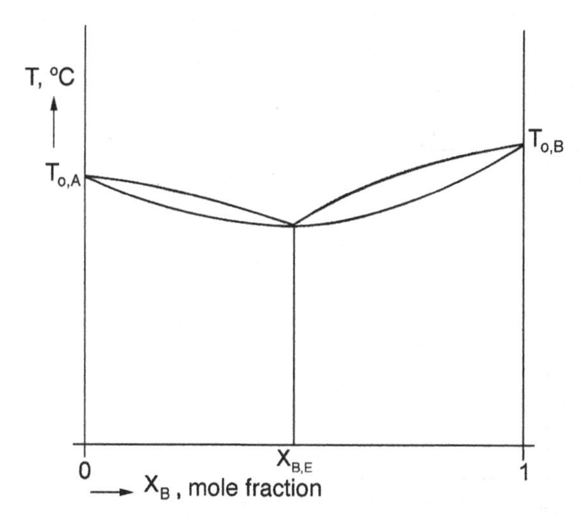

FIG. 7.5 Binary system with interaction between the two components.

whereas the composition of the solid changes according to the solidus. The lever rule is also applicable in this case. It is clear that purification by means of melt crystallization is more difficult than for eutectic systems. The system *p*-dibromobenzene/*p*-dichlorobenzene is an example of this category.

Continuous solid solubility is met when the pure components are not interacting. The phase diagram in Fig. 7.5 shows a minimum indicating an interaction between the two components. Except for the pure components and the minimum melting mixture, all compositions show a temperature range of melting or solidification. This case is relatively common. The binary mixture *d*-carvoxime/*l*-carvoxime possesses a maximum instead of a minimum. According to Sloan and McGhie (1988), this is the only example of such behavior reported to date.

7.3.3. Partial Miscibility

The reader is referred to Sloan and McGhie (1988).

7.4. CONSTITUTIONAL SUPERCOOLING

Possible temperature and concentration profiles for layer crystallization are shown in Fig. 7.6. The heat of fusion is transferred through the thermal boundary layer in the melt, the crystalline layer, and the heat exchanger wall. The concentration profile shown is the one for the impurity. In the mass transfer boundary layer, the impurity concentration increases because the

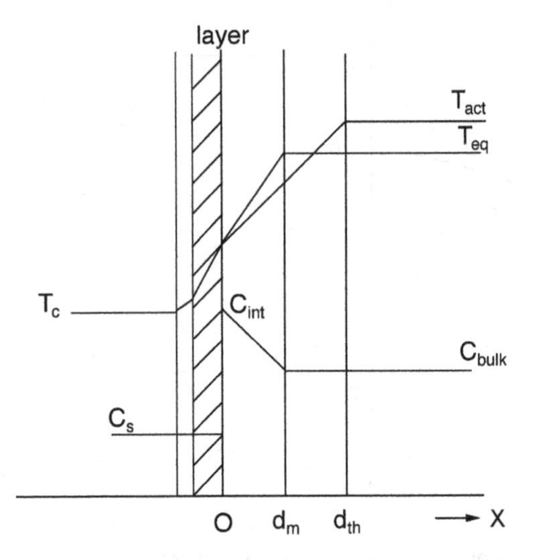

FIG. 7.6 Temperature and concentration profiles for layer crystallization.

transport of the desired component proceeds by means of diffusion. The ratio between the impurity concentration in the solid at the interface and the impurity concentration in the melt at the interface is the distribution coefficient. Usually, the thermal boundary layer thickness exceeds the diffusion boundary layer thickness. An approximate expression is:

$$\frac{d_{\mathrm{m}}}{d_{\mathrm{th}}} = \left(\frac{D}{a}\right)^{1/3} \tag{7.2}$$

The mass transfer by diffusion is outdone by convective mass transfer at a certain distance from the layer. However, molecular heat conduction is outdone by turbulent heat transfer at a larger distance from the layer.

Example

It is requested to calculate the ratio of these layer thicknesses for the melt crystallization of naphthalene. The material is almost pure.

Naphthalene physical properties

T_o = 80.4°C
ρ_l = 975 kg · m^{-3} (82°C)
c_l = 1710 J · kg^{-1} · K^{-1} (85°C)
λ_l = 0.135 W · m^{-1} · K^{-1} (89°C)
μ = 0.96 · 10^{-3} N · sec · m^{-2} (80.4°C)
D = 10^{-9} m^2 · sec^{-1} (assumption)

$$\frac{d_m}{d_{th}} = \left(\frac{D}{a}\right)^{1/3} = \left(\frac{c_l \rho_l D}{\lambda_l}\right)^{1/3} = \left(\frac{1710 \cdot 975 \cdot 10^{-9}}{0.135}\right)^{1/3} = 0.23$$

When $D = 10^{-8}\ \mathrm{m^2 \cdot sec^{-1}}$, the ratio is 0.50.
The conclusion is that d_{th} is two to four times larger than d_m.

Going back to Fig. 7.6, the essence is that the process temperature starts to drop at a given distance from the layer and that the impurity concentration is not yet affected at this point. The line $T_{eq} = f(X)$ indicates the temperature at which crystalline material is formed from the melt having an impurity concentration $c(X)$. It is clear that a zone exists in which the melt is supersaturated. This phenomenon is called constitutional supercooling and it may lead to dendritic growth. An incidental surface protrusion enters a zone of higher supersaturation and tends to grow faster than the original surface. This is illustrated in Fig. 7.7. By the same token, small dips in the crystal surface tend to grow slower than the surface. From Fig. 7.7, it can be understood that mother liquor can be occluded between the branches of the dendritic structures. If carried out at too high crystallization rates, layer crystallization can result in relatively impure products because of constitutional supercooling. If the layer crystallization is carried out at a relatively low crystallization rate, the impurity concentration at the interface is relatively low and T_{eq} can be lower than T_{act} for all X-values. In that case, there will be no constitutional supercooling.

Suspension crystallization is always carried out under conditions of constitutional supercooling (Fig. 7.8). The reason is that the heat of fusion is now transferred from the growing crystal to the melt. However, at suspension crystallization, the linear crystal growth rate is much smaller than at layer

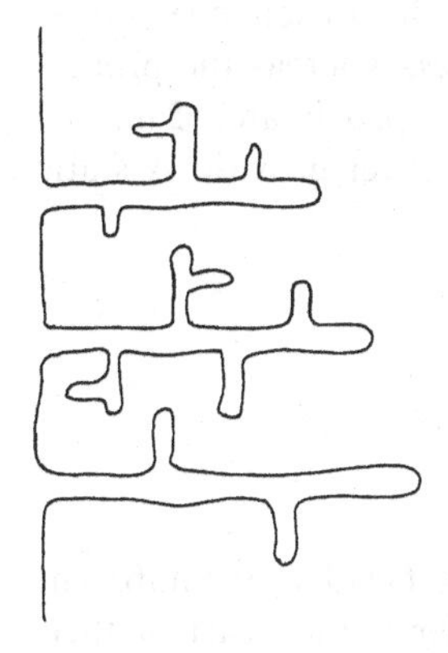

FIG. 7.7 Dendritic growth due to constitutional supercooling.

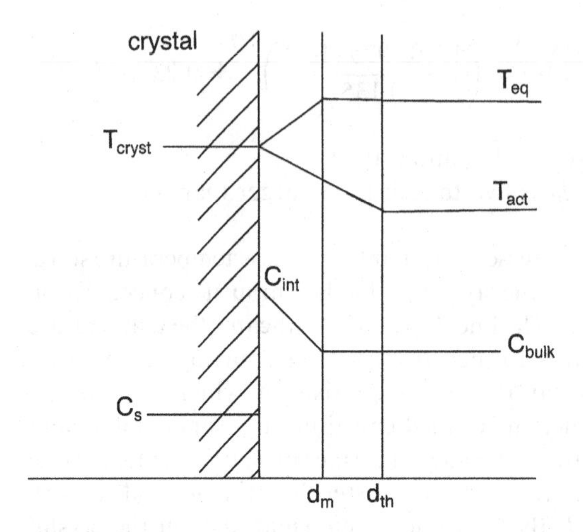

FIG. 7.8 Constitutional supercooling at suspension crystallization.

crystallization because of a larger surface area. The latter fact keeps the constitutional cooling low enough to avoid dendritic growth (see Sec. 7.6.2 for suspension growth characteristics).

7.5. LAYER GROWTH

7.5.1. Introductory Remarks

As described in Sec. 7.2, the first known application concerned the purification of monochloroacetic acid by BASF. Today, the PROABD refiner manufactured by BEFS Prokem (Mulhouse, France) is well known. The crystallizer is a rectangular closed vessel with finned heat exchanger tubes. The cooling/heating medium flows through the tubes, whereas the process material is in the vessel. The height of a crystallizer is approximately 2 m. The PROABD refiner is a piece of equipment for static layer growth. A static operation cycle comprises five phases:

- Filling with liquid feedstock
- Crystallizing part of the feed
- Draining the remaining liquid
- Sweating by temperature increase
- Melting the crystalline layer.

Sulzer Chemtec (Buchs, Switzerland) is the market leader for equipment for dynamic layer growth. A vertical heat exchanger is the heart of their system. The process material flows as a falling film through the tubes, whereas the cooling/heating medium flows as a falling film on the outside of the tubes. The BASF system as marketed by BEFS Prokem comprises the pumping of melt through heat exchanger tubes completely filled with melt. Likewise, the cooling/heating medium is pumped through the medium-side of the heat exchanger while this side is filled with liquid.

For both static layer growth and dynamic layer growth, scale-up is performed by putting in parallel a number of crystallization tubes such as the one tested at pilot plant experiments. The crystallization proceeds identically if the flows and the temperature control are identical.

A three-stage operation is shown in Fig. 7.9. The rectangles drawn indicate process stages. From left to right: recovery stage, first purification stage, and second purification stage. It is assumed that there is one crystallizer in this plant. As already indicated, after each crystallization, the temperature of the crystallized material is raised. This causes the liquefaction of part of the solid material and this liquid is run off. The effect of this operation is an additional purification. The liquid removed from the crystalline layer is called reflux. The process liquid used as feed for the first purification stage (feed stage) is made up of four different parts:

- The product from the previous process step (feed from, e.g., the distillation)
- The product from the recovery stage (stripping stage)

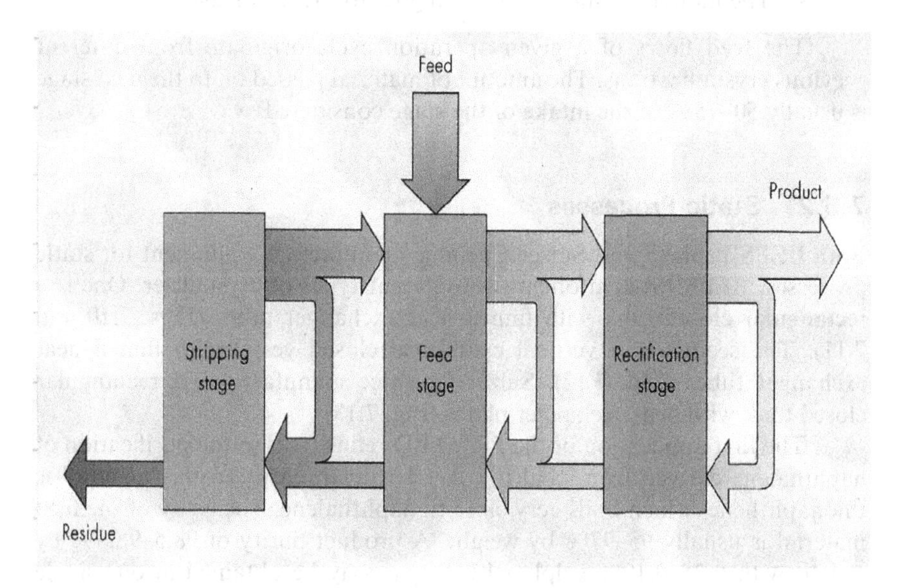

FIG. 7.9 Layer crystallization: a three-stage operation. (Courtesy of Sulzer Chemtec, Buchs, Switzerland.)

- The reflux of the first purification stage (internal recycle)
- The mother liquor from the second purification stage (rectification stage).

The crystallization of the first purification stage results in crystalline material and a remaining mother liquor. This step is a separation step. The mother liquor is part of the feed of the recovery stage. On raising the temperature of the crystalline material, reflux is obtained. This step is also a separation step. This reflux is part of the feed the next time a first purification stage is executed. The molten crystalline material is part of the feed for the second purification step. The other part is the reflux of the second purification step. The feed of the second purification stage is partly crystallized and the remaining liquid is passed on to the first purification stage. After removing the reflux of the second purification stage, the crystalline material can be molten. The latter step results in the product of the plant. The feed of the recovery stage is composed of its reflux and the mother liquor from the first purification stage. The feed of the recovery stage is also partly crystallized and the remaining liquid is rejected. The reflux of the recovery stage is recycled internally and the molten crystalline material is passed on to the first purification stage.

For each stage can be stated:

- The mother liquor goes to the previous stage.
- The reflux stays in the stage.
- The molten crystalline material goes to the next stage.

The feed flows of a given operation cycle originate from different previous crystallizations. The amount of material passed on to the next stage is usually 50–75% of the intake of the stage considered.

7.5.2. Static Processes

Both BEFS Prokem and Sulzer Chemtec manufacture equipment for static processes. BEFS Prokem offers two different types of crystallizer. One is a rectangular closed tank with finned heat exchanger tubes (Figs. 7.10 and 7.11). The second is a vertical cylindrical closed vessel with finned heat exchanger tubes (Fig. 7.12). Sulzer Chemtec manufactures a rectangular closed tank with heat exchanger plates (Fig. 7.13).

The first application of the PROABD refiner was in the purification of naphthalene derived from coal tar. A relevant impurity in this material is thionaphthene, which boils very close to naphthalene. The assay of the feed material is usually 95–97% by weight. A product purity of 98.5–98.8% by weight with 0.25–0.19% sulphur by weight may be obtained in one single operation (Arkenbout, 1995).

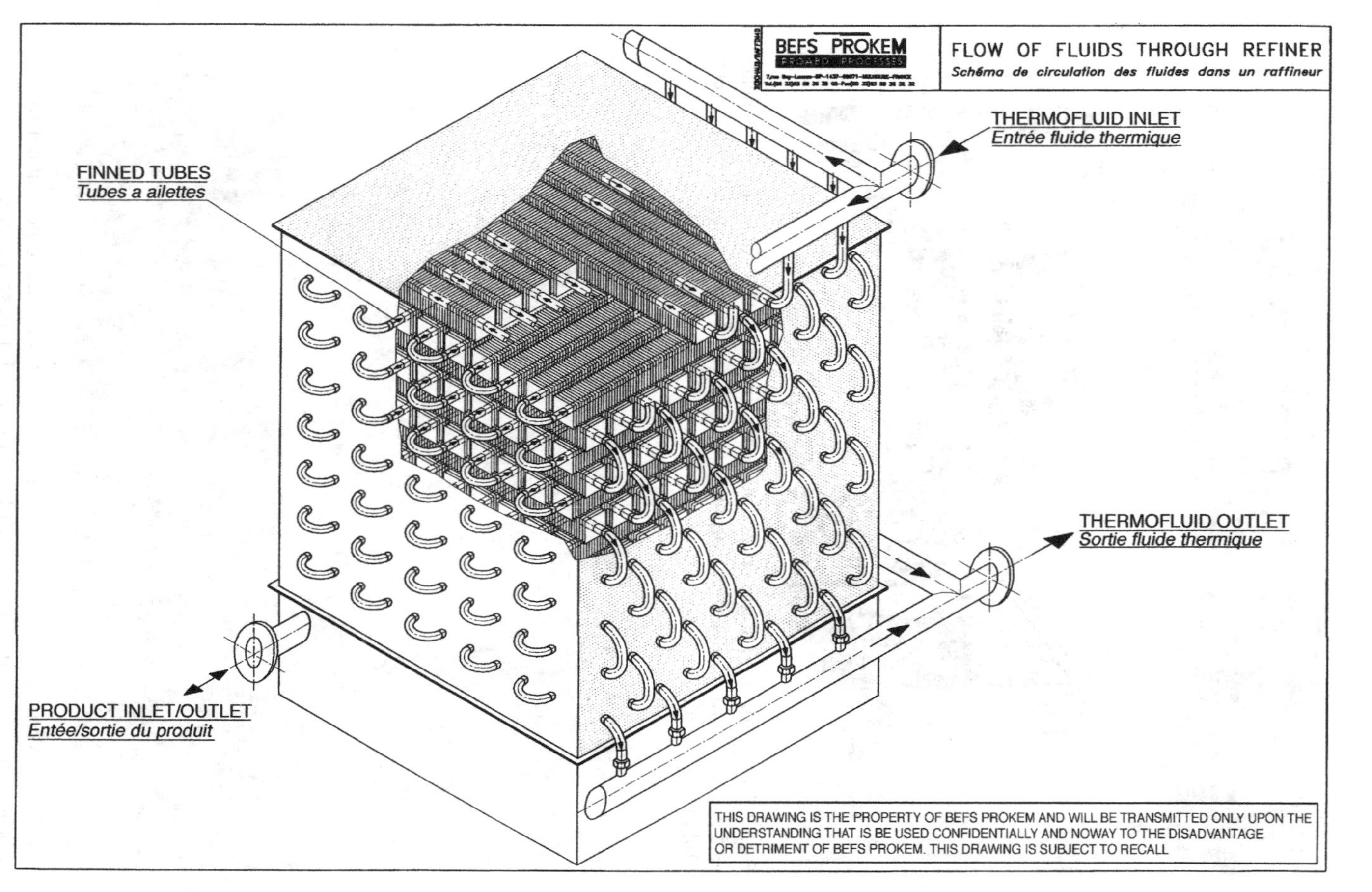

FIG. 7.10 Horizontal static crystallizer. (Courtesy of BEFS Prokem, Mulhouse, France.)

FIG. 7.11 Horizontal static crystallizers. Maximum 14,000 tons of naphthalene-99.5 per annum for China Steel Corporation in Taiwan. (Courtesy of BEFS Prokem, Mulhouse, France.)

FIG. 7.12 Vertical static crystallizer. 1300 tons of anthracene per annum for Elf Atochem in France. (Courtesy of BEFS Prokem, Mulhouse, France.)

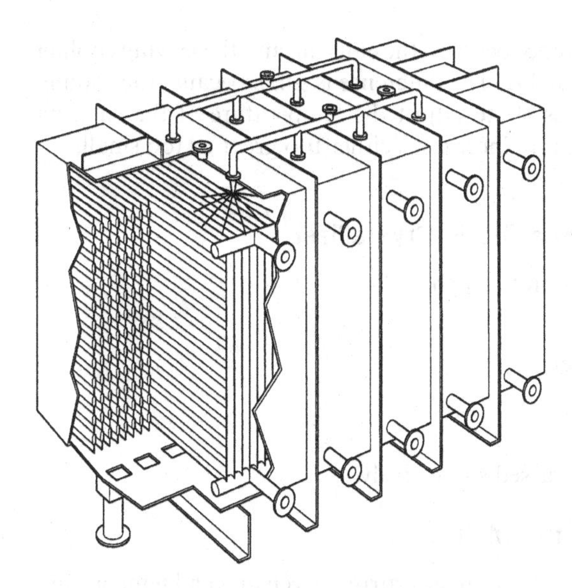

FIG. 7.13 Horizontal static crystallizer. (Courtesy of Sulzer Chemtec, Buchs, Switzerland.)

Constitutional supercooling is inevitable in the stagnant melt, and unstable growth morphologies are hence bound to occur. Networks of dendritic structures fill up the available space between the heat exchanger tubes. Impure residual mother liquor is entrapped in the crystal layer. This necessitates long drainage and sweating times. The crystallization is slow to achieve the desired purification. Typically, the temperature difference between the melt and the cooling water is initially 1–2 K. As the crystalline layer develops, this temperature difference is raised and may finally be 10–20 K. A typical crystallization lasts 20 hr and a cycle may last 24 hr.

The main resistance to heat transfer at static layer crystallization is in the growing layer. For example, the heat transfer coefficient of a naphthalene layer having a thickness of 10 mm is $0.135/0.01 = 13.5\ W \cdot m^{-2} \cdot K^{-1}$. This coefficient is much lower than the heat transfer coefficient of the tube wall and the medium-side coefficient. It is also much lower than the coefficient for heat transfer from the melt to the layer; here a typical figure for natural convection is $100\ W \cdot m^{-2} \cdot K^{-1}$ (Sec. 7.5.3).

Static layer crystallization processes are primarily used for small capacities and very high contaminant concentrations. The process is usually less sensitive to crystallization properties. For a fixed cooling water flow, the temperature rise is a measure of the degree of crystallization. The temperature rise is used for process control.

The temperature difference between the entering and the leaving cooling water flow is approximately 0.5 K. The difference is small because the cooling water flow is large. It is possible to adjust the temperature of the entering cooling water if the temperature increase is either too great or too small.

7.5.3. Heat Transfer in a Static Crystallizer

Heat is transferred by diffusion through:

- The melt
- The crystalline layer
- The tube wall
- The cooling medium.

These steps will be discussed successively.

Heat Transfer Through the Melt

The contents of a static crystallizer are not stirred. Even at small temperature differences, heat transfer by natural convection is an important mechanism. The correlation applicable for vertical surfaces is (Beek et al., 1999):

$$Nu = 0.55(Gr \cdot Pr)^{1/4} \left(10^3 < Gr \cdot Pr < 10^8\right) \tag{7.3}$$

In Nu, L is the length coordinate of the vertical surface. The correlation is also approximately correct for horizontal tubes. In that case, L should be replaced by D:

$$Nu = \frac{\alpha_o D}{\lambda_l}; \quad Pr = \frac{\nu}{a} = \frac{\mu c_l}{\lambda_l}; \quad Gr = \frac{L^3 g \Delta\rho_l}{\nu^2 \rho_l}$$

Example

A static crystallizer is equipped with tubes having an outside diameter of 3 cm. There are fins on the tubes having a diameter of 10 cm. The distance between the fins is 1 cm.

Naphthalene is crystallized. It is requested to calculate the process-side heat transfer coefficient for the horizontal tubes. For the calculation of the Grashof number, a ΔT of 1 K is assumed:

Naphthalene physical properties

$T_o = 80.4°C$
$\rho_l = 975\ kg \cdot m^{-3}$ (82°C)
$\Delta\rho_l = 0.8\ kg \cdot m^{-3}$ for 1 K
$c_l = 1710\ J \cdot kg^{-1} \cdot K^{-1}$ (85°C)

$\lambda_l = 0.135\ \text{W} \cdot \text{m}^{-1} \cdot \text{K}^{-1}\ (89°\text{C})$
$\mu = 0.96 \cdot 10^{-3}\ \text{N} \cdot \text{sec} \cdot \text{m}^{-2}\ (80.4°\text{C})$
$D = 10^{-9}\ \text{m}^2 \cdot \text{sec}^{-1}$ (assumption)

$$Pr = \frac{\mu c_l}{\lambda_l} = \frac{0.96 \cdot 10^{-3} \cdot 1710}{0.135} = 12.16$$

$$Gr = \frac{L^3 g \Delta\rho_l}{\nu^2 \rho_l} = \frac{0.03^3 \cdot 9.81 \cdot 0.8 \cdot 975}{(0.96 \cdot 10^{-3})^2} = 224,174$$

$$Nu = 0.55\ (12.16 \cdot 224{,}174)^{1/4} = 22.3$$
$$\alpha_o = 100.4\ \text{W} \cdot \text{m}^{-2} \cdot \text{K}^{-1}$$

Heat Transfer Through the Crystalline Layer

The heat transfer coefficient of a naphthalene layer having a thickness of 10 mm is $0.37/0.01 = 37\ \text{W} \cdot \text{m}^{-2} \cdot \text{K}^{-1}$.

Heat Transfer Through the Tube Wall

The heat transfer coefficient of a tube wall having a thickness of 2 mm is:

$$\frac{15}{0.002} = 7500\ \text{W} \cdot \text{m}^{-2} \cdot \text{K}^{-1}$$

Heat Transfer to the Cooling Medium

A correlation for forced convection through tubes is (Beek et al., 1999):

$$Nu = 0.027 \cdot Re^{0.80} \cdot Pr^{1/3} \left(\frac{\mu}{\mu_w}\right)^{1/7} \tag{7.4}$$

$$\left(2 \cdot 10^3 < Re < 10^5;\ Pr \geq 0.7\right)$$

Example

The inner diameter of the tubes of the previous example is 28 mm and water having a temperature of 78 °C is pumped through these tubes at a rate of $2\ \text{m} \cdot \text{sec}^{-1}$. It is required to calculate the coefficient for the heat transfer from the wall to the flowing water.

Water physical properties

$\rho_l = 973\ \text{kg} \cdot \text{m}^{-3}$
$c_l = 4198\ \text{J} \cdot \text{kg}^{-1} \cdot \text{K}^{-1}$
$\lambda_l = 0.675\ \text{W} \cdot \text{m}^{-1} \cdot \text{K}^{-1}$
$\mu = 0.365 \cdot 10^{-3}\ \text{N} \cdot \text{sec} \cdot \text{m}^{-2}$

The viscosity correction $(\mu/\mu_w)^{1/7}$ is not used because the temperature difference is small.

$$Nu = 0.027\left(\frac{973 \cdot 2 \cdot 0.028}{0.365 \cdot 10^{-3}}\right)^{0.80}\left(\frac{0.365 \cdot 10^{-3} \cdot 4198}{0.675}\right)^{1/3} = 488.8$$

$$\alpha_c = \frac{488.8 \cdot 0.675}{0.028} = 11,784 \text{ W} \cdot \text{m}^{-2} \cdot \text{K}^{-1}$$

Overall Heat Transfer

The overall heat transfer coefficient for the four steps can now be calculated as follows:

$$\frac{1}{U} = \frac{1}{100.4} + \frac{1}{37} + \frac{1}{7500} + \frac{1}{11,784} = 0.0372$$

$$U = 26.9 \text{ W} \cdot \text{m}^{-2} \cdot \text{K}^{-1}$$

The overall heat transfer is strongly determined by the thermal conductivity of the naphthalene layer. The presence of the fins is not taken into account in the calculations in this section.

7.5.4. Dynamic Processes

Sulzer Chemtec is a market leader with its batchwise dynamic falling-film system. A typical crystallization plant is shown in Fig. 7.14. The crystallizer consists of a series of vertical tubes. The liquid to be crystallized is fed at the top of the tubes by means of the circulation pump. The melt flows down the inner wall of the tube as a film and solidifies on the wall as the outside of the tubes is cooled. Cooling is accomplished by a cooling medium flowing down the outer wall of the tube also as a film. The melt flowing into the collector tank is recycled to the top of the crystallizer. The desired ratio between the amount of material crystallized and the amount of mother liquor remaining can be checked by means of the level or the mass in the collector tank. The latter can be checked by means of load cells.

The crystallization step is usually followed by a sweating step. The final step is the melting of the crystalline layer. The crystallization plant shown in Fig. 7.14 is suitable for the execution of three purification stages. To illustrate this, Stage 2 will be described in some detail.

The middle vessel (the Stage 2 vessel) contains the mother liquor of the third stage of the previous cycle. It also contains the reflux of the second stage of the previous cycle. Feed material is added to this vessel.

The contents of the middle vessel are drained to the collecting tank. This melt is recirculated through the crystallizer and the heating loop is started. The crystalline material that remained in the crystallizer

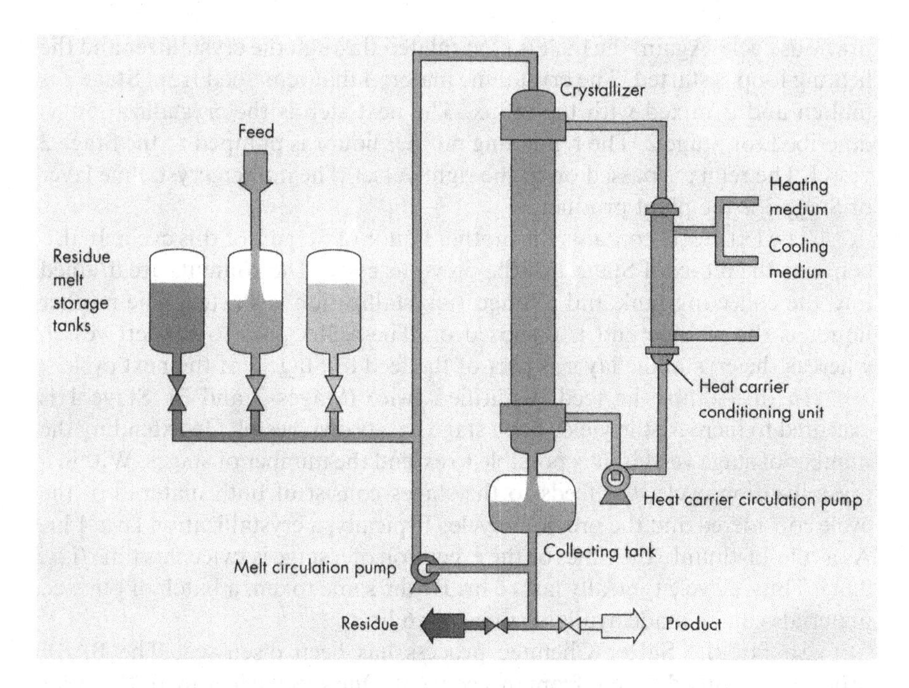

FIG. 7.14 Falling film crystallizer. (Courtesy of Sulzer Chemtec, Buchs, Switzerland.)

from Stage 1 is molten and added to the feed. The mixed material is the feed for Stage 2.

The crystallization of Stage 2 is started by circulating the melt and activating the cooling system. During crystallization, the coolant temperature is gradually lowered. With regard to heat transfer, the increasing temperature difference compensates for the increasing resistance to heat transfer of the growing layer. The temperature of the circulating melt falls slightly because the assay of the melt decreases. The process is stopped when the required amount of material has been crystallized.

The mother liquor in the collecting tank is pumped to the left vessel. The left vessel is the Stage 1 vessel.

The crystalline material in the crystallizer is partially molten by starting the heating loop. The amount of reflux is checked in the collecting vessel.

The reflux is pumped to the middle vessel; it is part of the feed for Stage 2 of the next cycle.

Stage 3 is started by draining the contents of the right vessel (the Stage 3 vessel) to the collecting tank. The contents are the reflux of Stage 3 of the previous cycle. Again, the melt is recirculated through the crystallizer and the heating loop is started. The crystalline material that remained from Stage 2 is molten and is mixed with the reflux. The next step is the crystallization as described for Stage 2. The remaining mother liquor is pumped to the Stage 2 vessel. The reflux is passed on to the right vessel. The molten crystalline layer of Stage 3 is the plant product.

The left vessel contains the mother liquor of Stage 2 of this cycle. It also contains the reflux of Stage 1 of the previous cycle. The contents are drained into the collecting tank and a Stage 1 crystallization is started. The mother liquor is the residue and is disposed of. The reflux goes to the left vessel, whereas the crystalline layer is part of the feed of Stage 2 of the next cycle.

In this setup, the feed is purified twice (Stages 2 and 3). Stage 1 is executed to increase the yield. Each stage has its own vessel. On extending the number of stage vessels, it is possible to extend the number of stages. Within a crystallization cycle, the feeds to the stages consist of both material of the cycle considered and the previous cycle. Typically, a crystallization lasts 1 hr. As a rule-of-thumb, the time for the execution of a stage is twice this time (i.e., 2 hr). Thus, a cycle typically lasts 6 hr. By the same token, a batch of purified materials can be made available once per 6 hr.

So far, the Sulzer Chemtec process has been discussed. The BASF process is slightly different from this concept. One specific feature of the latter process is that the melt is circulated through vertical heat-exchanging tubes completely filled with melt. The medium-side of the vertical crystallizer is also completely filled with (usually) cold or warm water. In the Sulzer Chemtec process, a careful distribution of both melt and medium at the top of the crystallizer is required because they flow down as films. The advantage is that the crystallization conditions are the same for each tube. The distribution of the melt and the medium is less critical in the BASF process. The (small) drawback is that the crystallization conditions are not exactly the same for each tube. At the melt side, the conditions are equal. At the medium side, the remark made is applicable. However, the drawback is not very important because the resistance to heat transfer is small at the medium side.

Example

Naphthalene is crystallized in a vertical crystallizer with tubes. The heat transfer coefficient at the process side is 1500 $W \cdot m^{-2} \cdot K^{-1}$. The heat transfer coefficient at the medium side is 5000 $W \cdot m^{-2} \cdot K^{-1}$. Solid naphthalene's thermal conductivity is 0.37 $W \cdot m^{-1} \cdot K^{-1}$ at 38°C (Groot Wassink, 1976).

Calculate the various resistance percentages for a layer thickness of 10 mm:

$$\frac{1}{5000} + \frac{10^{-2}}{0.37} + \frac{1}{1500} = 2 \cdot 10^{-4} + 2.7 \cdot 10^{-2} + 6.7 \cdot 10^{-4} = 27.9 \cdot 10^{-4}$$

$$\text{Medium side}: \frac{2 \cdot 10^{-4}}{27.9 \cdot 10^{-3}} \cdot 100 = 0.7\%$$

$$\text{Layer}: \frac{2.7 \cdot 10^{-2}}{27.9 \cdot 10^{-3}} \cdot 100 = 96.9\%$$

$$\text{Process side}: \frac{6.7 \cdot 10^{-4}}{27.9 \cdot 10^{-3}} \cdot 100 = 2.4\%$$

Rittner and Steiner (1985) discuss a typical example of a dynamic process. It deals with the purification of benzoic acid. The purified acid is used for food and pharmaceutical purposes. The specifications concerning purity and especially color and smell are very stringent. The feed assay is 99.2% by weight and the assay of the product is 99.99% by weight. A Sulzer Chemtec falling-film crystallizer is used for this purification. Six recrystallizations in a series are carried out. The feed to the plant is part of the feed of Stage 2. At crystallizations, the melt temperature gradually increases from 110–115 °C to 121 °C in going from Stage 1 to Stage 6. At the start of each crystallization, the cooling medium is approximately 10 K cooler than the melt. The temperature of the cooling medium is gradually lowered by 15 K during the crystallization. Each crystallization lasts 25 min and a cycle lasts 265 min.

The melt film flowing down the inner tube wall has a thickness of 1–2 mm. This is accomplished by feeding 0.7 $1 \cdot sec^{-1}$ and per meter of tube circumference. The cooling medium is fed at a rate of 1.1 $1 \cdot sec^{-1}$ and per meter of tube circumference.

For the material balance, the following data are given. The plant produces 6000 tpa (tons per annum). The feed to Stage 2 is composed as follows:

- 4.7 t of benzoic acid having a purity of 99.2% by weight
- 2 t of mother liquor ex Stage 3 of the previous cycle
- 0.7 t of reflux ex Stage 2 of the previous cycle
- 3 t of molten crystals ex Stage 1 of the previous cycle.

These feed flows add up to 10.4 t. This amount is separated into three fractions:

- 6 t of crystals
- 0.7 t of reflux
- 3.7 t of mother liquor.

Thus, 57.7% by weight of the feed is crystallized. The waste acid contains 97% benzoic acid by weight. With this figure, it is possible to calculate the yield. Assume that X tons of the product leave the plant per ton of material. The relevant equation is: $0.992 \cdot 1 = 0.9999X + 0.97\,(1-X)$. It follows that $X = 0.736$. The benzoic acid yield is:

$$\frac{0.9999 \cdot 0.736}{0.992 \cdot 1} \cdot 100 = 74.2\%$$

Rittner and Steiner (1985) also give data concerning equipment sizes:

- 450 tubes
- Tube length of 12 m
- Tube diameter of 50 mm.

In this type of plant, the only moving parts are pumps and valves. Pilot plant tests are performed with one tube. The results can be scaled-up by designing a crystallizer having a number of parallel tubes.

7.5.5. Design Method for a Falling Film Crystallizer

The time needed to achieve a certain layer thickness of the crystalline material will be dealt with. Starting points:

- The crystallization process is determined by the heat transfer.
- The crystallization of the pure compound is taken into account only.
- Purities will not be calculated.
- The temperatures of the falling melt film and the liquid/solid interface of the crystalline layer are equal.

Medium-Side Heat Transfer

Beek et al. (1999) give relationships for the heat transfer to a falling film:

$$\alpha_c = 2.05\left(\frac{4\rho_l v\delta}{\mu}\right)^{-1/3}\left(\frac{g\rho_l^2\lambda_l^3}{\mu}\right)^{1/3} \tag{7.5}$$

for

$$\frac{4\rho_l v\delta}{\mu} < 1600$$

If $4\rho_l v\delta/\mu > 3200$, the film is turbulent and the heat transfer can be described by the equation:

$$\alpha_c = 0.0087\left(\frac{4\rho_l v\delta}{\mu}\right)^{0.4}\left(\frac{g\rho_l^2\lambda_l^3}{\mu^2}\right)^{1/3}\left(\frac{v}{a}\right)^{1/3} \tag{7.6}$$

Tube Wall

The heat transfer coefficient is λ_w/d_w.

Growing Layer

The solidification process is depicted in Fig. 7.15. First, heat conduction through a hollow cylinder is discussed (Fig. 7.16). The radius of the outer cylinder is R_2, whereas the radius of the inner cylinder is R_1. Heat is conducted from the inner cylinder to the outer cylinder. The temperature of the inner cylinder is T_1, whereas the outer cylinder's temperature is T_2. The heat flow from the inner part of the hollow cylinder to the outside is constant. Therefore, the heat flow through any thin layer having a thickness dr can be expressed as follows:

$$Q = -\frac{\lambda_s}{dr} 2\pi r \cdot dT \quad W \cdot m^{-1}$$

This is the heat flow per meter of tube length.
Integrating the differential equation yields:

$$Q = \frac{2\pi\lambda_s(T_1 - T_2)}{{}^e\log \frac{R_2}{R_1}} \quad W \cdot m^{-1} \tag{7.7}$$

This equation can also be written as:

$$Q = \frac{\lambda_s}{R_2 - R_1} \cdot 2\pi R_m(T_1 - T_2) \quad W \cdot m^{-1}$$

R_m is the logarithmic mean of R_1 and R_2:

$$R_m = \frac{R_2 - R_1}{{}^e\log \frac{R_2}{R_1}}$$

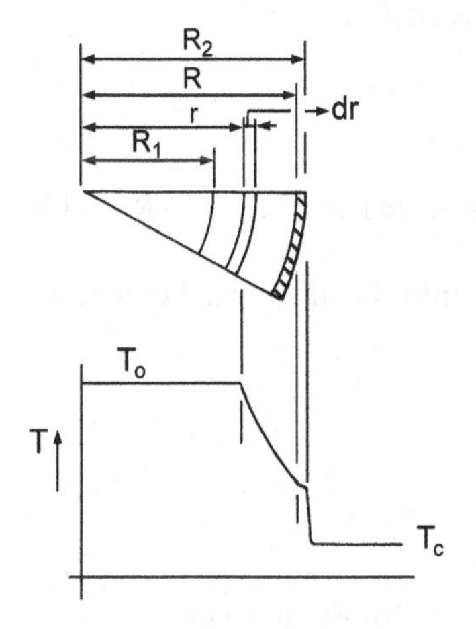

FIG. 7.15 Growing of a layer in a dynamic crystallizer.

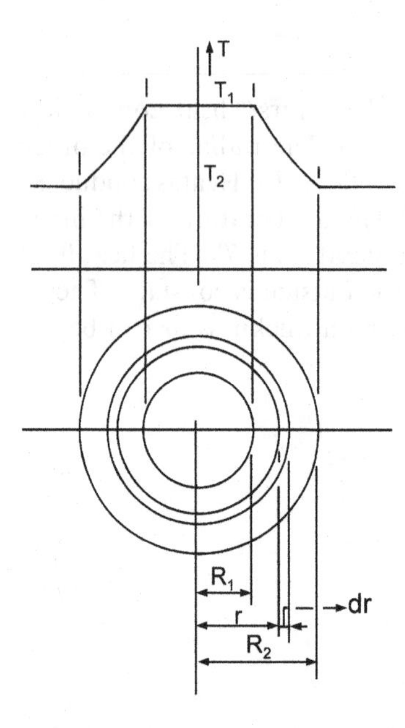

FIG. 7.16 Heat conduction through a hollow cylinder.

Second, the solidification of a cylinder will be discussed. The differential equation describing the case depicted in Fig. 7.15 is:

$$k2\pi r(T_o - T_c)\mathrm{d}t = -\rho_s i 2\pi r \cdot \mathrm{d}r \quad \mathrm{J \cdot m^{-1}}.$$

This equation neglects the enthalpy effect of the cooling hollow cylinder. The enthalpy effect will be discussed after the completion of the treatment of the present case. k is defined by the following equation:

$$\frac{1}{kr} = \frac{1}{\alpha_k R} + \frac{{}^{e}\log \dfrac{R}{r}}{\lambda_s}$$

Integration of the differential equation between $r = R$ and $r = R_1$ yields the time for solidification.

First, the expression for kr is substituted into the differential equation. This yields:

$$\mathrm{d}t = -\frac{\rho_s i r\left(\lambda_s + \alpha_k R \ {}^{e}\log \dfrac{R}{r}\right)\mathrm{d}r}{\alpha_k R \lambda_s (T_o - T_c)}$$

Reworking this equation leads to:

$$\mathrm{d}t = -\frac{\rho_s i}{\alpha_k R \lambda_s (T_o - T_c)}(\lambda_s r + \alpha_k R r \ {}^{e}\log R - \alpha_k R r \ {}^{e}\log r)\mathrm{d}r$$

The integration of the first two terms of the right-hand side (RHS) of the equation is straightforward. The integration of the third term of the RHS proceeds as follows:

$$\int r\,{}^{e}\log r \cdot dr = \frac{1}{2}\int {}^{e}\log r \cdot dr^2 = \frac{1}{2}\left\{({}^{e}\log r)r^2 - \int r^2 d\,{}^{e}\log r\right\}$$

The integration of the third part of the RHS of the differential equation is now also straightforward on realizing that $d\,{}^{e}\log r = dr/r$. The time required for the crystallization of a hollow cylinder having a wall thickness of $(R - R_1)$ can be calculated as follows:

$$t = \frac{\rho_s i}{2\alpha_k \lambda_s R(T_o - T_c)}\left\{\left(\lambda_s + \frac{\alpha_k R}{2}\right)(R^2 - R_1^2) + \alpha_k R R_1^2\,{}^{e}\log\frac{R_1}{R}\right\} \text{ sec} \tag{7.8}$$

The time required for the complete crystallization of a cylinder having a radius of R can be expressed by the equation:

$$t = \frac{\rho_s i R}{4\alpha_k (T_o - T_c)}\left(2 + \frac{\alpha_k R}{\lambda_s}\right) \tag{7.9}$$

On defining three dimensionless numbers, a simple expression can be obtained for the latter case.

These three numbers are (see also Sec. 5.5):

$Fo = \frac{at}{R^2}$ [the Fourier number, $a = \lambda_s/(c_s \rho_s)$]

$Ph = \frac{i}{c_s(T_o - T_c)}$ [the Phase Transfer number, (latent heat)/(sensible heat)]

$Bi = \frac{\alpha_k R}{\lambda_s} = \left(\frac{R}{\lambda_s}\right) \Big/ \left(\frac{1}{\alpha_k}\right)$ (the Biot number, the ratio of the Internal resistance to heat transfer to the external resistance)

The simple expression is:

$$Fo = \frac{Ph}{2}\left(\frac{1}{Bi} + \frac{1}{2}\right) \tag{7.10}$$

Equations (7.8), (7.9), and (7.10) were derived by neglecting the specific heat of the solid material. The equations tend to become complicated when this effect is taken into account. For a Phase Transfer number of 5, the calculated time can be multiplied by 1.15. For greater values, the correction factor becomes smaller and vice versa (e.g., see VDI, 1993). See also Chapter 2.

Example 1

Naphthalene is crystallized in a falling-film crystallizer. The inner and outer tube diameters are 100 and 106 mm ($R = 0.05$ m and $R_2 = 0.053$ m). The temperature of the cooling water is 70°C and the feed rate is 1.1 l · sec^{-1} and per meter of tube circumference.

It is required to calculate the time needed for the crystallization of a layer having a thickness of 18 mm ($R_1 = 0.032$ m).

Naphthalene Physical Properties

$T_o = 80.4°C$
$i = 148{,}100 \text{ J kg}^{-1}$
$\rho_s = 1150 \text{ kg} \cdot \text{m}^{-3}$ (60°C)
$c_s = 1440 \text{ J} \cdot \text{kg}^{-1} \cdot \text{K}^{-1}$ (51°C)
$\lambda_s = 0.37 \text{ W} \cdot \text{m}^{-1} \cdot \text{K}^{-1}$ (38°C) (Groot Wassink, 1976)

Water Physical Properties at 70°C

$\rho_l = 978 \text{ kg} \cdot \text{m}^{-3}$
$c_l = 4194 \text{ J} \cdot \text{kg}^{-1} \cdot \text{K}^{-1}$
$\lambda_l = 0.660 \text{ W} \cdot \text{m}^{-1} \cdot \text{K}^{-1}$
$\mu_l = 0.404 \cdot 10^{-3} \text{ N} \cdot \text{sec} \cdot \text{m}^{-2}$

Calculation of the medium side heat transfer coefficient

$$\rho_l v \delta = 978 \cdot 1.1 \cdot 10^{-3} = 1.08 \text{ kg} \cdot \text{m}^{-1} \cdot \text{sec}^{-1}$$

$$\frac{4 \cdot 1.08}{0.404 \cdot 10^{-3}} = 10{,}693$$

The film flow is turbulent:

$$\alpha_c = 0.0087(10{,}693)^{0.4} \left(\frac{9.81 \cdot 978^2 \cdot 0.660^3}{0.404^2 \cdot 10^{-6}} \right)^{1/3}$$

$$\times \left(\frac{0.404 \cdot 10^{-3} \cdot 4194}{0.660} \right)^{1/3} = 12{,}405 \quad \text{W} \cdot \text{m}^{-2} \cdot \text{K}^{-1}$$

Calculation of the heat transfer coefficient of the combination of the water film and the metal wall:

$$\frac{1}{\alpha_k R} = \frac{1}{\alpha_c R_2} + \frac{{}^e\log \dfrac{R_2}{R}}{\lambda_w}$$

$\lambda_w = 15.5\ \text{W} \cdot \text{m}^{-1} \cdot \text{K}^{-1}$ for stainless steel
$\alpha_k = 3788\ \text{W} \cdot \text{m}^{-2} \cdot \text{K}^{-1}$

Equation (7.8):

$$t = \frac{1150 \cdot 148{,}100}{2 \cdot 3788 \cdot 0.37 \cdot 0.05(80.4 - 70)} \left\{ \left(0.37 + \frac{3788 \cdot 0.05}{2}\right) \times (0.05^2 - 0.032^2) + 3788 \cdot 0.05 \cdot 0.032^2\ {}^e\log\frac{0.032}{0.05} \right\} = 6282 \text{ sec}$$

$$Ph = \frac{148{,}100}{1440(80.4 - 70)} = 9.9$$

Corrected time: $1.1 \cdot 6282 = 6910$ sec (almost 2 hr)

Example 2

The increasing resistance to heat transfer of the growing crystalline layer is often compensated by lowering the cooling water temperature. It is required to calculate the time required for the crystallization of the naphthalene layer of the previous example when the cooling water temperature falls linearly from 70°C to 50°C in 1 hr:

$$T_c = -\frac{t}{180} + 70$$

$$T_o - T_c = T_o + \frac{t}{180} - 70$$

The original differential equation now becomes:

$$\left(T_o + \frac{t}{180} - 70\right)dt = -\frac{\rho_s i}{\alpha_k R \lambda_s}(\lambda_s r + \alpha_k R r\,{}^e\log R - \alpha_k R r\,{}^e\log r)dr$$

It follows from the result of the previous example:

$$\frac{t^2}{360} + T_o t - 70t = 6282(80.4 - 70) \quad \text{and} \quad t = 3326 \text{ sec}$$

$1.1 \cdot 3326 = 3659$ sec.

For this example, a typical growth rate is $18 \cdot 10^{-3}/3659 = 4.92 \cdot 10^{-6}$ $\text{m} \cdot \text{sec}^{-1}$.

7.5.6. Laboratory Experiments

The "bottle test" is a simple technique for checking the suitability of the layer crystallization process for the purification of a compound. It comprises cooling down a sample in a glass bottle until it has partially frozen. The type of crystals grown, the ease of the separation of both fractions, the masses of the crystals and the melt, and the analyses of the fractions are relevant. Generally, at static layer crystallizations, the temperature differences are quite small. The temperature differences at the "bottle test" are usually not so small. Static layer crystallization hence shows a higher purification efficiency than the "bottle test." Dynamic layer crystallization is different from the test procedure because the melt flows turbulently through the tubes. However, the "bottle test" is a useful orientating test. Wintermantel (1986) proposed a 1–l stirred crystallization vessel. The vessel has a flat bottom that can be cooled. The cooling rate and the stirrer speed are two independent process variables. The crystallizer can be turned upside down to allow draining and sweating. This type of equipment is suitable for the investigation of dynamic layer crystallizations. BEFS Prokem uses glass tubes to study static layer crystallization at laboratory scale. The glass tubes are jacketed. A typical tube has a height of 2 m and an internal diameter of 40 mm. The tube is filled with melt and the crystallization process is started by circulating cooling water through the jacket. The cooling water temperature is gradually lowered. Crystallization, draining, sweating, and remelting occur successively. The cooling rate is the independent process variable.

7.5.7. Pilot Plant Tests

Usually, devices are used that contain one or several tubes having the same diameter as tubes that are used industrially. The key to the success of scaling-up is the ability to operate the tubes of an industrial crystallizer in an identical way. The main items are the distribution of the melt flows and the temperature control.

BEFS Prokem's pilot plant unit for static layer crystallization has a volume of 24.5 l. The unit is made of stainless steel; 15 basic elements are stacked one on top of the other. The total height is approximately 2 m. Each basic element consists of a chamber with a finned tube. Warm water or cooling water runs through the finned tubes. The finned tubes of the elements are connected so as to form one long spiral-shaped tube. In consecutive elements, the tubes are perpendicular to each other.

Arkenbout (1995) describes the BASF pilot equipment for dynamic layer crystallization. The crystallizer consists of two tubes having a diameter of 25 mm and a length of 5 m. The two tubes are operated in series. Wintermantel and Wellinghof (1991) report that laboratory data and pilot plant data are in line.

7.5.8. Comparison of Static and Dynamic Processes

It will be assumed that in both types of equipment, the mass transfer is determined by the heat transfer. It has been shown that the heat transfer in both static and dynamic equipment is strongly determined by the resistance to heat transfer of the crystalline layer. Typically, in a static process, the crystallization lasts approximately a day. A dynamic layer crystallization typically lasts an hour. There are three reasons for this difference:

- At static crystallizations, the temperature difference between the melt and the cooling medium is smaller than for dynamic crystallizations.
- The process-side heat transfer coefficient for static crystallizations is worse than the one for dynamic crystallizations.
- At static crystallizations, thicker layers are grown.

Jancic (1989) compared static and dynamic layer crystallizations of benzoic acid. For both types of crystallization, benzoic acid, having an assay of 99% by weight, was the feed. A single-step dynamic operation results in a product having an assay of 99.5% by weight in 2–3 hr. The static method results in a 99.9% product in 20 hr. Three consecutive dynamic steps result in a slightly better purity than the static step's purity. Static crystallizations are primarily used for small capacities and very high contaminant concentrations. Furthermore, static crystallization is not well suited for multistage duty because the long cycle times require large storage facilities. Dynamic crystallization is a better choice as the cycle times are much shorter. It is possible to combine static and dynamic layer crystallizations in one plant. Static crystallization can be used to produce nearly spent residues at low operation cost, whereas dynamic crystallization is suitable for the performance of ultra-purification at high throughputs.

Typically, a static crystallizer provides 100 m^2 heat exchanging area per cubic meter. This figure is approximately 20 for a dynamic crystallizer. The reason for this difference is the presence of fins in static crystallizers. Finally, it is remarked that static crystallization is usually less sensitive to crystallization properties.

7.5.9. Typical Application of a Static Process

BEFS Prokem carried out pilot plant experiments for Akzo Nobel. It was requested to purify a feed having an assay of 87.2% by weight. The desired product purity is 98.5% by weight minimum. The pure product has a melting point of 40 °C. The first step was the execution of laboratory experiments (Sec. 7.5.6). These were successful; the purification required two stages. Next, tests were carried out in the 25-l unit (Sec. 7.5.7). The unit could be filled with approximately 22 kg. One crystallization cycle lasted approximately 20 hr; the cooling water temperature was gradually lowered from 50°C to 30°C. A crystallization cycle comprises filling, crystallization, draining, sweating, and melting. The crystallization per se lasted 14 hr. Appendix A contains the material balances of three consecutive purification steps. The tests were carried out on a once-through basis; recycling was not practiced. The first two steps show the upgrading of the material; however, the third step hardly offers a bonus. This is in line with the fact that static layer crystallization is hardly or not suitable for ultrapurification. Concerning the compositions of the flows of the first two purification steps, conclusions can be drawn as follows:

- Reflux first stage $\neq$ feed first stage
- Mother liquor second stage = feed first stage
- Reflux second stage = feed second stage.

Furthermore, the reflux of the first stage is 15% of the feed of the first stage.

Because of these facts, the pilot plant tests on a once-through basis can be considered to be representative for an industrial operation with recycles.

Fig. 7.17 contains an approximate material balance for an industrial plant with an output of $1.00\ \mathrm{t \cdot hr^{-1}}$. The composition of the reject can be calculated from the compositions of the feed and the product:

$$1.35 \cdot 0.872 = 0.35X + 1.00 \cdot 0.987X \rightarrow X = 0.55$$

The yield is: $(1 \cdot 0.987 - 0.55 \cdot 0.35)/(0.872 \cdot 1.35)100 = 67.5\%$
Based on the first stage, the crystallizer volume should be:

$$\frac{40 \cdot 2.04}{0.9 \cdot 0.8} = 113.3\ \mathrm{m^3}$$

Explanation

Once per 40 hours, the intake is $40 \cdot 2.04$ t. The specific mass of the melt is 0.9 $\mathrm{t \cdot m^{-3}}$ and the crystallizer volume is occupied for 80%.

7.6. SUSPENSION GROWTH

7.6.1. Introduction

Layer growth was discussed in Sec. 7.5. Suspension melt crystallization is based on a different concept. The solid material crystallizes on crystals suspended in a melt. At the same time, nuclei are generated in the melt. This technique is comparable to solution suspension crystallization as applied to vacuum pan salt and other products. The main advantage over the two layer methods is the large specific crystal area. Typically, this parameter can be $10{,}000\ \mathrm{m^2 \cdot m^{-3}}$. Suspension growth characteristics are discussed in Sec. 7.6.2. Suspension melt crystallization has the same potential problems as suspension solution crystallization: the need to control the crystal size distribution, the

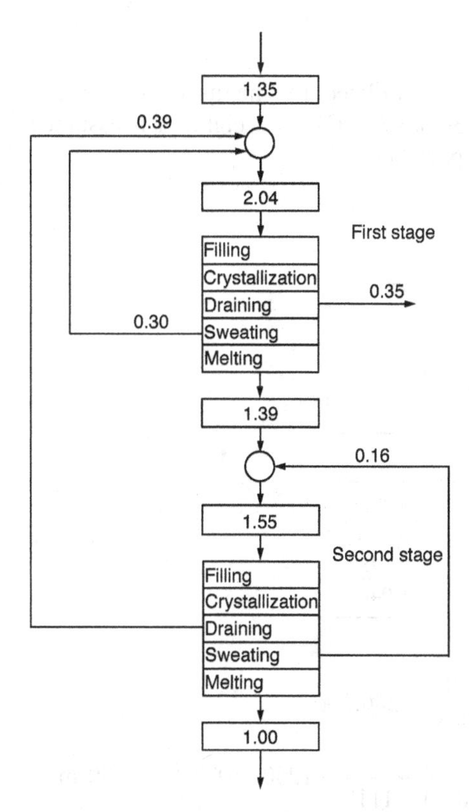

FIG. 7.17 Material balance for production of $1 \text{ t} \cdot \text{hr}^{-1}$.

handling of suspensions and melts, and the solid/liquid separation. Furthermore, the viscosity of melts is often higher than the viscosity of solutions.

Because of these aspects, with the exception of *p*-xylene and *p*-dichlorobenzene recovery, suspension processes are hardly used industrially for separating organics. The suspension crystallization of *p*-xylene is discussed as an example in Sec. 7.6.3.

The population density theory is a tool to control particle size distributions. It is discussed in Sec. 7.6.4.

7.6.2. Suspension Growth Characteristics

Compared to layer crystallization, the specific crystal area in square meters per cubic meter is 10–100 times larger for suspension growth. This will be illustrated by an example.

Example

An organic peroxide (OP) is crystallized by means of suspension crystallization. The slurry contains 30% solid OP by weight. It is assumed that the crystals are cubical 250-μm particles.

Physical data	
Melting point pure OP	40.0°C
Specific mass solid OP	$1110 \text{ kg} \cdot \text{m}^{-3}$
Specific mass liquid OP	$900 \text{ kg} \cdot \text{m}^{-3}$

Calculation			
	kg	$\text{kg} \cdot \text{m}^{-3}$	m^3
Solid OP	300	1110	0.270
Liquid OP	700 +	900	0.778 +
	1000		1.048

$$\text{Mass of crystals in 1 m}^3: \frac{300}{1.048} = 286.3 \text{ kg}$$

$$\text{Crystal area per m}^3: \frac{286.3}{(250 \cdot 10^{-6})^3 \cdot 1110} \cdot 6(250 \cdot 10^{-6})^2 = 6190 \text{ m}^2$$

In a falling-film crystallizer having 0.1-m diameter tubes, the initial crystal area per cubic meter tube volume is $\pi \cdot 0.1/\{(\pi/4) \cdot 0.1^2\} = 40 \text{ m}^2$.

For a falling-film crystallizer, the area per cubic meter crystallizer volume is much smaller than 40 m^2.

A typical figure for static layer crystallizers is $100 \text{ m}^2 \cdot \text{m}^{-3}$ crystallizer volume.

Because of the large crystal area, the growth rate is low at suspension growth. This will also be illustrated by an example.

Example

OP is crystallized by means of suspension crystallization. One cubic meter of the slurry of the previous example is stirred in a crystallizer.

Crystallizer data

Diameter = 1 m
Cylindrical height = 1.2 m

Jacket height = 1 m
Temperature difference crystallizer contents/jacket = 20 K
Heat transfer coefficient = 300 W · m^{-2} · K^{-1}

Physical data
Heat of fusion of OP = 75,000 J · kg^{-1}

Calculation
Jacket area $\pi 1 \cdot 1$ = 3.14 m^2
Q = 300 · 3.14 · 20 = 18,840 W
Production: 18,840/75,000 = 0.2512 kg · sec^{-1}

It is now possible to calculate the linear growth rate. Again, it is assumed that the crystals are cubical 250-μm particles. $6190 \cdot r \cdot 1110 = 0.2515$ kg · sec^{-1} and $r = 3.66 \cdot 10^{-8}$ m · sec^{-1}. Compare this linear growth rate to the linear growth rate calculated for dynamic layer crystallization in Sec. 7.5.5: $4.92 \cdot 10^{-6}$ m · sec^{-1}.

There is always constitutional supercooling at suspension growth. This was explained in Sec. 7.4. Still, it is possible to obtain pure crystals because the growth rate is relatively low.

The small specific mass difference between liquid and solid favors mild agitation. In solution crystallization, this difference is usually much greater. Again, this will be illustrated by an example.

Example

Compare the solution crystallization of NaCl to the melt suspension crystallization of OP.

	ρ_s (kg · m^{-3})	ρ_l (kg · m^{-3})
NaCl	2160	1250
OP	1110	900

Mild agitation favors the formation of relatively coarse crystals because the crystals are not abraded. However, mild agitation does not favor secondary nucleation. Thus, secondary nuclei are often introduced by a wall scraper. A further influence is the viscosity of the liquid. For example, the viscosity of liquid OP at 50°C is 9 cP.

The viscosity of brine is approximately 1 cP. The relatively high viscosity of organic melts also favors mild agitation. However, mixing becomes more difficult when the viscosity is high. Moreover, filtration or centrifugation becomes less effective at high viscosity.

At suspension melt crystallization, the temperature affects the slurry density strongly. However, the heat of fusion acts as a buffer. In principle, the temperature cannot change when both solid and liquid are present (e.g., the temperature of an ice/water system cannot change until all the water has become ice). To solidify the liquid, the heat of fusion must be removed. This requires time.

Example

A total of 0.2512 kg of OP is produced per second in the second example of this section. There are 286.3 kg of crystals in the crystallizer. In 1 min, 5.3% of the suspended materials is produced.

At suspension crystallization, slurries must be handled. This aspect is the main disadvantage of the process.

Stirrers are required in the crystallizers to avoid sedimentation. Often, these stirrers also act as wall scrapers because crusts tend to grow on the jacketed walls of the crystallizer.

Special pumps are required to handle the suspensions. Wash columns are often used for solid/liquid separation. In these columns, mechanical transport devices are often required because the density difference between solid and liquid is small. The agitator construction can be complicated and great forces are sometimes required to overcome incrustation. Furthermore, agitators can also suffer themselves from incrustations and cause backmixing. Generally, in wash columns, highly precise temperature control is necessary to avoid encrustation.

Slurries can be handled on a large scale only after extensive laboratory and pilot plant testing.

As a rule, the crystallization process consists of three steps: crystallization, mechanical separation, and purification. Quite often, the separation and purification are performed simultaneously. Solid/liquid separation can be carried out by means of centrifuges, belt filters, or wash columns. Wash columns are preferred often because the temperature can be controlled.

7.6.3. The Melt Crystallization of *p*-Xylene—A Typical Example

Introduction

p-Xylene is mainly oxidized to terephthalic acid, which can be esterified to dimethyl terephthalate. The latter component is a precursor for polyesters.

The C_8 aromatic mixture obtained at the processing of crude oil is the most important source of *p*-xylene. A typical crude xylene feed composition is given (Takegami and Meyer, 1996):

	% by weight	Melting point (°C)	Melting point (K)
Ethyl benzene	1.5	−94.825	178.175
p-Xylene	21.5	13.413	286.413
m-Xylene	50.5	−47.722	225.278
o-Xylene	26.5	−25.032	247.968

Two successful methods to isolate *p*-xylene are molecular sieve adsorption and suspension crystallization. A characteristic of adsorption processes such as the Parex process is the high recovery per single pass.

Prior to suspension crystallization, *o*-xylene can be separated from the mixture by means of distillation:

	o-Xylene	*m*-Xylene	*p*-Xylene	Ethyl benzene
Boiling point at 1 bara [°C]	144.4	139.1	138.4	136.2

This distillation will raise the concentration of *p*-xylene in the crude xylene feed, which will raise the recovery per single pass of the crystallization process.

On cooling the typical crude xylene feed, *p*-xylene is the first component to come out of the solution. The temperature where the second component comes out of the solution can be calculated. This temperature is called the first eutectic. It is not possible to operate a crystallization unit below this temperature if a high purity for *p*-xylene is required. The calculation of the first eutectic will be dealt with in two steps. First, the binary system *m*-xylene/*p*-xylene is discussed.

The Van 't Hoff expression is discussed in Sec. 7.3. McKay et al. (1966) do not recommend the use of this equation when *p*-xylene recoveries are to be calculated. A better accuracy can be obtained by the application of an equation involving the use of cryoscopic constants. These constants are made available by the Thermodynamics Research Center (TRC, 1996). The equation is:

$$-{}^{e}\log X_{A} = A(T_{o,A} - T)\{1 + B(T_{o,A} - T)\} \tag{7.11}$$

The cryoscopic constants for the components of the crude xylene feed are listed in Table 7.1. Equation (7.11) enables the calculation of the melting

TABLE 7.1 Component Data

Component	Cryoscopic constant A	Cryoscopic constant B
Ethyl benzene	0.03471	0.0029
p-Xylene	0.02599	0.0028
m-Xylene	0.02741	0.0027
o-Xylene	0.02659	0.0030

point depression substance A experiences in a solution. Solid A starts to come out of the solution when a solution containing a mole fraction X_A of substance A is cooled to temperature T. Because the solution is assumed to be ideal, the nature of the other components in the solution is irrelevant. A similar equation is applicable for substance B. The temperature and the composition of the eutectic point can be found by combining the equations for substances A and B.

Example 1

To calculate the temperature and the composition of the eutectic point of the system *m*-xylene/*p*-xylene.

At the eutectic point:

$$-{}^{e}\log X_A = 0.02741(225.278 - T_E)\{1 + 0.0027(225.278 - T_E)\} \text{ and}$$

$$-{}^{e}\log(1 - X_A) = 0.02599(286.413 - T_E)\{1 + 0.0028(286.413 - T_E)\}$$

It follows: $X_A = 0.86$ and $T_E = 220$ K (−53°C).

Fig. 7.18 shows the phase diagram of the binary system *m*-xylene/*p*-xylene.

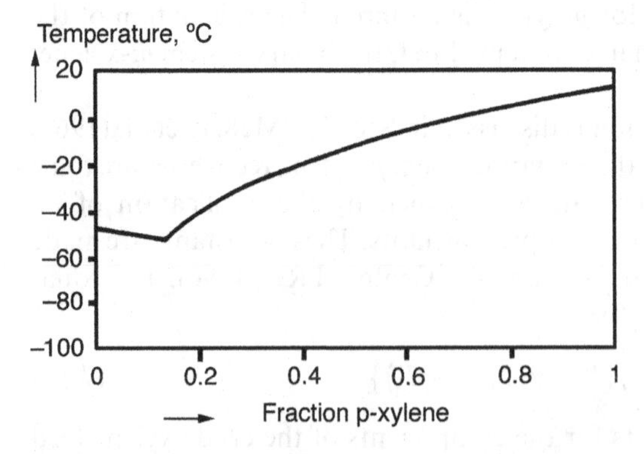

FIG. 7.18 Phase diagram for *m*-xylene/*p*-xylene.

Now, the first eutectic for the crude xylene feed will be determined. The first step is to calculate the temperature at which *p*-xylene starts to crystallize. This temperature is a function of the mole fraction of *p*-xylene only. The next step is the assumption of a temperature for the first eutectic. *p*-Xylene crystallizes only on cooling a crude xylene feed from the temperature at which *p*-xylene starts to crystallize to the temperature of the first eutectic. The mole fraction of *p*-xylene in the liquid phase at the assumed temperature of the first eutectic can be calculated by using Eq. (7.11). The mole fractions of the other components in the liquid can be calculated from the feed composition. It is also possible to calculate the saturated mole fraction for each component other than *p*-xylene at the assumed temperature of the first eutectic. For one of the components other than *p*-xylene, the saturated mole fraction should be equal to the actual mole fraction. For the other components other than *p*-xylene, the saturated mole fractions should be greater than the actual mole fractions. Finding the temperature of the first eutectic may involve some iteration.

Example 2

This example is quoted from Takegami and Meyer (1996). It is summarized in Table 7.2. In this example, *o*-xylene has not been separated from the crude xylene feed by distillation.

The first column contains the names of the components. The second column gives the feed composition. The third column contains the temperatures at which each component would start to crystallize if the other components did not crystallize. For example, the temperature of −189°C is the point at which ethyl benzene would start to crystallize from a solution where its concentration was 1.5% and the other components in the mixture did not crystallize. The temperatures in the third column can be calculated by

TABLE 7.2 Multicomponent Eutectic Calculation

			At −65°C	
Component	Feed composition	Freezing point [°C]	Liquid composition	Saturated composition
Ethylbenzene	1.5	−189	1.75	258.57
p-Xylene	21.5	−38	8.38	8.38
m-Xylene	50.5	−71	58.94	61.19
o-Xylene	26.5	−69	30.93	30.57

The compositions are in percent by weight or mole percent.
The saturated composition calculated for ethyl benzene is above 100% because the system temperature is above the pure component freezing point.
The slurry density at −65°C is 21.5 − (50.5/58.94)8.38 = 14.3% by weight.

using Eq. (7.11). A temperature of −65°C has been assumed as the temperature of the first eutectic. At −65°C, the liquid phase contains 8.38% *p*-xylene by weight. This percentage can be found in the fourth column. The other percentages can be calculated directly from the feed composition in the second column. Of course, these percentages are valid only if crystallization of components other than *p*-xylene did not occur.

Finally, the fifth column contains the concentrations at which the components start to crystallize at −65°C. The values in the fifth column should be greater than or equal to the values in the fourth column. Because only *o*-xylene and *p*-xylene have saturated compositions at or below the liquid composition, *o*-xylene is the second component to come out of solution (after *p*-xylene) and −65°C is the temperature of the first eutectic.

It is now possible to calculate the yield of *p*-xylene per pass.

	kg feed	kg reject
Ethyl benzene	1.5	1.5
p-Xylene	21.5	7.2
m-Xylene	50.5	50.5
o-Xylene	26.5 +	26.5 +
	100.0	85.7

$$\text{Yield}: \frac{21.5 - 7.2}{21.5} \cdot 100 = 66.5\%$$

Furthermore, when a process produces pure *p*-xylene from a 21.5% feed with a reject liquid of 8.38%, 85.7% of the feed material would have to be recycled back to an isomerization unit and returned to the crystallizer. The first eutectic is a major limitation.

In this approach, the cryoscopic constants are not a function of the liquid composition and therefore the calculated eutectic is only an approximation.

The calculated equilibria data can be checked experimentally by determining the temperature of the first eutectic. An example is quoted from McKay et al. (1966). The composition of McKay et al.'s mixture and the composition of the liquid phase as the mixture was cooled down are shown in Table 7.3. The samples were analyzed by infrared spectroscopy. These data show that *p*-xylene starts to crystallize between −65 and −84°F and that the first eutectic was reached at −97°F (−71.7°C). At this temperature, *o*-xylene also starts to come out of the solution. The calculated value is −93.1°F (−69.5°C). Mole percents instead of percents by volume are assumed.

TABLE 7.3 Phase Equilibria for Mixed Xylenes

Feed:			
Ethyl benzene	26.2		
p-Xylene	10.8		
m-Xylene	38.9		
o-Xylene	22.8		
Toluene	1.3		
Temperature (°F)	*m*-Xylene	*o*-Xylene	*m*-Xylene/*o*-xylene ratio
−65	39.4	23.0	1.71
−84	43.1	25.1	1.72
−96	44.5	25.9	1.69
−97	44.3	26.0	1.70
−98	46.7	22.2	2.10
−100	47.0	21.0	2.24

The composition is given in percent by volume.
Source: McKay et al. (1966).

The successful separation of *p*-xylene is favored by several aspects. First, *p*-xylene has a much higher freezing point than its isomers. The melting points of *p*-xylene and its isomers have been reviewed in the beginning of this section. Second, the melt has a low viscosity. Typically, the melt viscosity figures in a *p*-xylene suspension crystallization plant are smaller than 10 cP. Third, the nucleation and growth characteristics favor the formation of relatively large, well-shaped crystals. The crystals have the form of small beams having the largest dimension of hundreds of microns (McKay et al., 1966).

Solid/liquid separation is routinely carried out by means of centrifugation and has also been demonstrated in wash columns. Several suspension crystallization processes for the separation of *p*-xylene are commercially available. The TSK CCCC system will be discussed next. TSK stands for Tsukishima Kikai (Tokyo, Japan), whereas CCCC is an acronym for continuous countercurrent crystallization.

The TSK CCCC System for *p*-Xylene

The system is depicted in Fig. 7.19. In this setup, there are three cooling crystallizers and a purifier (also called wash column). The clear feed enters the second crystallizer (T2). Crystal slurries are stirred in the three crystallizers. In principle, the solid material is pure *p*-xylene. The process temperature of crystallizer T1 is approximately −65°C. In Example 2 in this section, *o*-xylene is the second component to come out of the solution at −65°C. The process

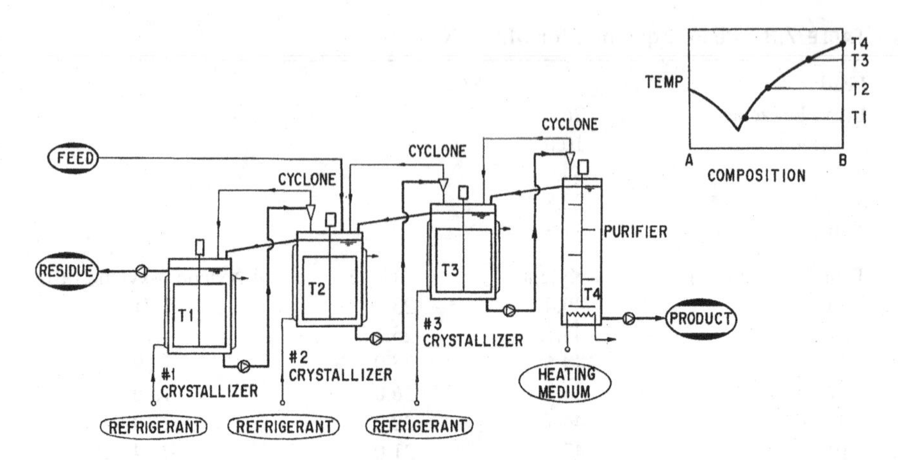

FIG. 7.19 Process flow diagram of TSK's CCCC system. (Courtesy of Tsukishima Kikai, Tokyo, Japan.)

temperature at the bottom of the purifier is 14°C (i.e., the melting point of pure *p*-xylene). The temperatures of the other two crystallizers are in the range of −65°C to 14°C. Based on the feed composition in Example 2 of this section, 85.7 kg flow back as residue per 100 kg feed. The product flow would thus be 14.3 kg per 100 kg feed.

The *p*-xylene crystals are transported from left to right, whereas "clear" liquid flows from right to left. The liquid is depleted of *p*-xylene at the latter process. Liquid/solid separation in the plant is accomplished by means of hydrocyclones. The heat of crystallization is removed by means of indirect heat exchange (cooling by flashing propane). The crystallizer agitator is a rotating draft tube (Fig. 7.20). Spring-loaded scrapers are mounted on the outside of the draft tube. The scrapers also exert a pumping action. Impellers are also mounted at the inside of the draft tube.

The melter at the bottom of the purifier liquifies all of the arriving crystals. A portion of the liquid flows upward as reflux. The crystals are purified while flowing downward by the countercurrent washing of this reflux. The remaining liquid is drawn off as product having a purity of up to 99.99%. In this setup, centrifuges are not used.

The purifier column contains an agitator, which rotates very slowly. This keeps the crystals evenly distributed over the cross section of the purifier and prevents channeling of the reflux.

TSK has built *p*-xylene plants with capacities up to 70,000 tons per annum.

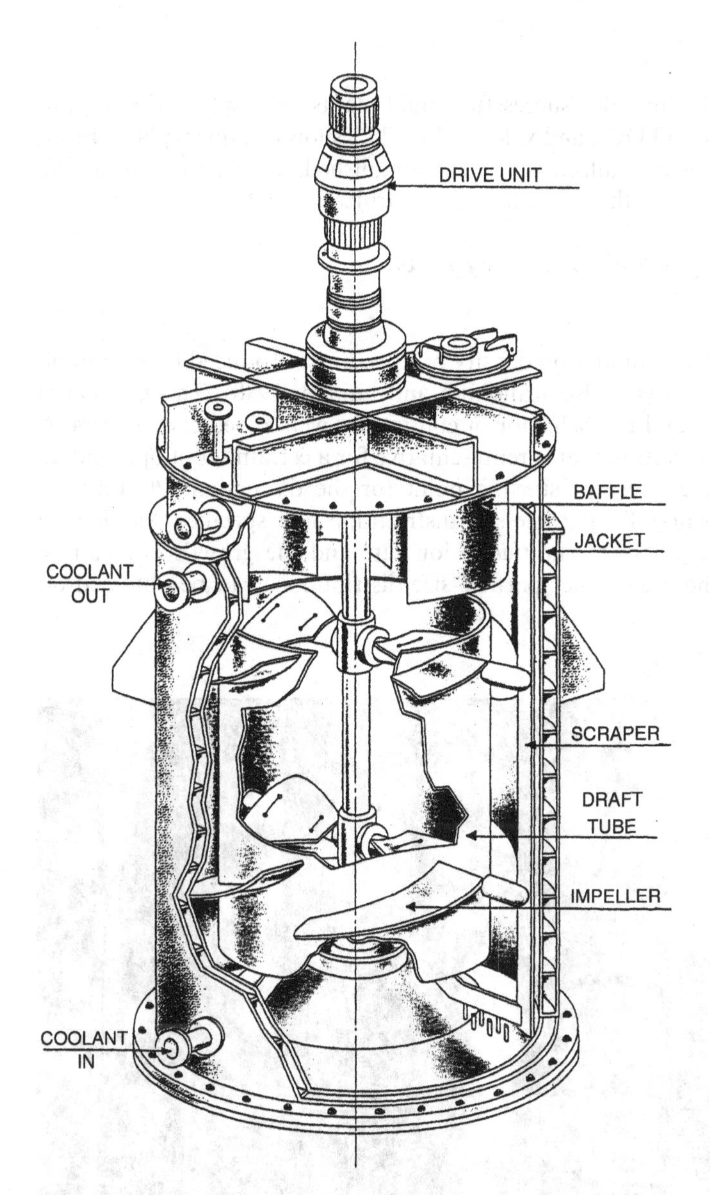

FIG. 7.20 Crystallizer of TSK's CCCC system. (Courtesy of Tsukishima Kikai, Tokyo, Japan.)

Final Note

Tsukishima Kikai has also successfully built plants for bisphenol A, terephthaloyl dichloride (TDC), and xylenol. Fig. 7.21 shows a typical plant. From left to right, four crystallizers can be distinguished. The last one bears the number 32B. Next to this crystallizer is a purification column.

7.6.4. The Population Density Balance

Introduction

The concept of the population density balance was introduced by Randolph and Larson (1962) (see also Randolph and Larson, 1988). This concept is useful to understand the behavior of continuous well-mixed crystallizers. A particle size distribution that is representative for a certain set of operational conditions is a convenient starting point for the discussion. Plotting the recalculated results of a particle size distribution in a specified way ideally results in a straight line. The nucleation rate and the growth rate can be derived from the plot. Other particle size distributions can be obtained on

FIG. 7.21 Melt crystallization plant. (Courtesy of Tsukishima Kikai, Tokyo, Japan.)

varying the residence time while keeping the other operational conditions constant. Each particle size distribution leads to a combination of nucleation rate and growth rate. Thus, sets of nucleation rates and growth rates are obtained. Often, on plotting the logarithm of the nucleation rate vs. the logarithm of the growth rate, a straight line can be drawn. This result can be used to predict the behavior of a crystallizer by interpolation or extrapolation.

It is also possible to vary the slurry density while keeping the other operational variables constant. Again, combinations of nucleation rate and growth rate are obtained and can be plotted similarly.

Finally, a single-variable study can be made of the power input due to agitation.

If the recalculated results of a particle size distribution do not lead to a straight line, physical causes for this phenomenon can be looked for.

The Population Density Plot

Typically, a particle size distribution is presented as the mass in grams per liter of slurry in a number of crystal classes. Each crystal class has an upper and a lower boundary (e.g., 200 and 100 μm). As a first step, the number of crystals in each class per liter of slurry is calculated from the mass of crystals in each class. For each class, this is done by dividing the mass of crystals by $\rho_s k_v L_{av}^3$. In the latter expression, k_v is the shape factor; k_v is one for cubes and $\pi/6$ for spheres. It is now possible to make a cumulative plot of N, the number of crystals per liter of slurry, as a function of the crystal size L. The derivative of this function dN/dL is called the population density n. For each L, n represents the slope of the curve $N = f(L)$. The concept of population density is comparable to, for example, probability density or IQ density.

Approximations of n at L_{av} can be obtained for each crystal class by dividing the number of crystals by the width of the crystal class. The unit of the population density is per micrometer per liter. It is remarked that there will be more crystals per liter of slurry in the size range 0–100 μm than in the size range 100–200 μm because crystals from the smaller size range either grow into the larger size range or leave the crystallizer. By the same token, the number of crystals in the size range 200–300 μm is smaller than the number of crystals in the size range 100–200 μm and so on.

It will now be shown that, for a Continuous Mixed Suspension Mixed Product Removal (CMSMPR) crystallizer, the plot of the natural logarithm of the population density as a function of the crystal size is a straight line. The formal proof will be preceded by a discussion of the behavior of a Continuous Stirred Tank Reactor (CSTR) to which a tracer is added. A CSTR is depicted in Fig. 7.22. The reactor's volume is V whereas ϕ_v flows through the vessel. An amount of tracer equal to $C_0 V$ is added at $t = 0$. Hence, the initial concentra-

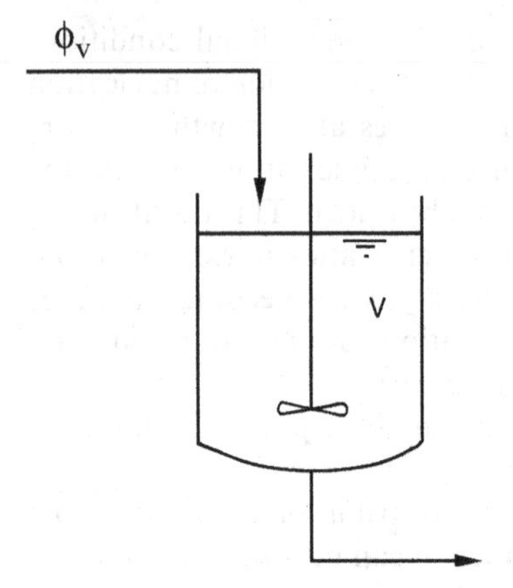

FIG. 7.22 A continuous stirred tank reactor.

tion in the reactor is $C_0V/V = C_0$. In a small time dt, the concentration in the CSTR falls with dC. A material balance:

$$-V\mathrm{d}C = C\phi_V\mathrm{d}t$$

Rearranging shows that the rate of decrease of the concentration is proportional to the concentration:

$$-V\frac{\mathrm{d}C}{\mathrm{d}t} = \phi_V C$$

It follows:

$$-\int_0^t \frac{\phi_V}{V}\mathrm{d}t = \int_{C_0}^{C} \frac{\mathrm{d}C}{C}$$

V/ϕ_V is equal to the residence time τ.
Integration yields $C = C_0 e^{-t/\tau}$.
The corresponding relationship for CMSMPR crystallization is:

$$n = n_0 e^{-t/\tau} = n_0 e^{-L/(r\tau)} \tag{7.12}$$

$t = L/r$ is the time the crystal has resided in the crystallizer. L is the crystal size and r is the rate of growth.

The analogy between the expressions for the concentration and the population density is clear. The expression for the concentration represents the decay of the tracer concentration as a function of time. Each concentration is unambiguously coupled to a certain lapse of time. Each crystal having size L has resided for $t = L/r$ min in the crystallizer. Thus, each crystal size is also unambiguously coupled to a certain lapse of time. By the same token, this statement can also be made for the population density.

The formal proof will now be given. A CMSMPR crystallizer is depicted in Fig. 7.23. The conditions are:

- The feed does not contain crystals.
- The crystallizer contents are mixed well.
- The crystal growth rate is independent of the size.
- Crystals do not agglomerate.
- Crystals do not break.

A numbers balance (population balance) for crystals in a size range ΔL is stated per second and per liter of slurry:

$$Vrn_1 = Vrn_2 + \phi_v \bar{n} \Delta L$$

where Vrn_1 is the number of crystals growing into the size range. Vrn_2 is the number of crystals growing out of the size range and $\phi_v \bar{n} \Delta L$ is the number of crystals in the size range removed from the crystallizer. It follows:

$$\frac{Vr(n_2 - n_1)}{\Delta L} + \phi_v \bar{n} = 0 \quad \text{and} \quad \frac{Vr\mathrm{d}n}{\mathrm{d}L} + \phi_v n = 0.$$

Again, the "rate of decrease" $\mathrm{d}n/\mathrm{d}L$ is proportional to n. This leads to an expression containing a natural logarithm. Rearranging and integrating leads to:

$$\int_{n_o}^{n} \frac{\mathrm{d}n}{n} = -\int_0^L \frac{\mathrm{d}L}{r\tau} \quad \text{and} \quad n = n_o \mathrm{e}^{-L(r\tau)} \tag{7.12}$$

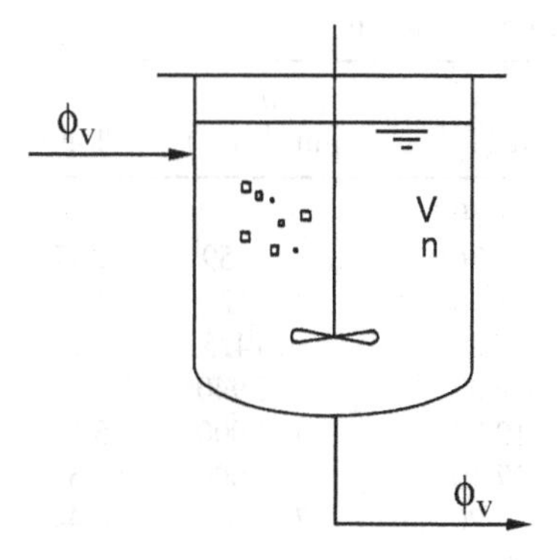

FIG. 7.23 A "Continuous Mixed Suspension Mixed Product Removal" crystallizer.

Example

Larson and Randolph (1969) give an example of a size distribution from CMSMPR crystallization. Their data are reproduced in Table 7.4. One liter of slurry contains 111.85 g of crystals. The masses of crystals in grams in nine size ranges are stated in the table. As a first step, the number of crystals in each class per liter of slurry is calculated from the mass of crystals in each class. Each mass of crystals is divided by $\rho_s k_v L_{av}{}^3$. $\rho_s = 1.77 \cdot 10^{-12}$ g $\cdot$ μm^{-3} whereas k_v, the shape factor, is 1. This means that the particles are cubes. The population density n for each crystal class is obtained by dividing the number of crystals in each crystal class by ΔL, the width of the size range. The population density is expressed per micrometer and per liter. $^{10}\log n$ is plotted as a function of L in Fig. 7.24.

This plot is a straight line but for two points representing only 2.5 g of material. The rate of growth of the crystals can be calculated from the slope of the line. On progressing from 100 to 200 μm ($\Delta L = 100$ μm), the population density has fallen by a factor of 100. Because $^{10}\log 100 = 2$, the relevant equation is $1/(2.303 \cdot r \cdot \tau) = 2/100$. For $\tau = 15$ min, $r = 1.45$ μm $\cdot$ min^{-1} ($2.4 \cdot 10^{-8}$ m $\cdot$ sec^{-1}). Note that the rate of growth is in two directions, so the rate of growth of one face of a cube is 0.725 μm $\cdot$ min^{-1}.

The nucleation rate is $4.8 \cdot 10^7 \cdot 1.45 = 6.96 \cdot 10^7$ min$^{-1} \cdot$ l^{-1}. Both the rate of growth and the nucleation rate have been obtained by reading the population density plot. The values for n_o and r must also satisfy an equation for the cumulative mass distribution. The expression representing the latter variable is $M_T = k_v \rho_s \int_0^L nL^3 dL$ up to size L.

In the limit:

$$M_T = 6 k_v \rho_s n_o (r\tau)^4 \text{ g} \cdot \text{l}^{-1} \tag{7.13}$$

TABLE 7.4 Size Distribution from CMSMPR Crystallization

Standard U.S. sieve	L_{av} (μm)	ΔL (μm)	$\rho_s k_v L_{av}{}^3 \Delta L$ (g $\cdot$ μm)	w (g $\cdot$ l^{-1})	n (μm$^{-1} \cdot$ l^{-1})	$^{10}\log n$
+35				1.30		
−35 + 40	479	66	$1.28 \cdot 10^{-2}$	0.76	59	1.77
−40 + 50	374	145	$1.34 \cdot 10^{2}$	1.70	127	2.10
−50 + 70	257	89	$2.68 \cdot 10^{-3}$	3.82	1425	3.15
−70 + 100	177	71	$6.98 \cdot 10^{-4}$	9.03	12,940	4.11
−100 + 140	128	26	$9.65 \cdot 10^{-5}$	12.72	132,000	5.12
−140 + 200	98	34	$5.65 \cdot 10^{-5}$	27.10	480,000	5.68
−200 + 325	63	37	$1.64 \cdot 10^{-5}$	44.70	2,730,000	6.44
−325				10.72 +		
				111.85		

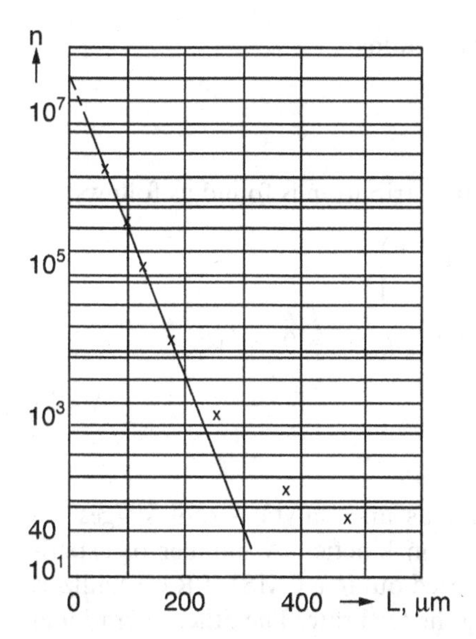

FIG. 7.24 $^{10}\log n$ as a function of L.

The values already found for n_o and r must satisfy this equation. Check:

$$M_T = 6 \cdot 1 \cdot 1.77 \cdot 10^{-12} \cdot 4.8 \cdot 10^7 \cdot (21.7)^4$$
$$= 113.0\ \text{g} \cdot \text{l}^{-1} \text{ and } 113.0 \approx 111.85.$$

The Dominant Size

The dominant size is the size at which the mass fraction distribution reaches a maximum. This parameter can be found by setting the derivative of the mass fraction distribution equal to zero. First, an expression for the mass fraction distribution is derived.

The mass of an individual particle is given as $m_1 = \rho_s k_v L^3$. The crystal mass per unit volume of slurry in a small size range dL is equal to the mass of one crystal times the number of crystals per unit volume of slurry dN in size range dL:

$$dm = \rho_s k_v L^3 dN \text{ and } dm = \rho_s k_v L^3 n dL$$

The total particle mass per liter of slurry is $M_T = 6\rho_s k_v n_o (r_T)^4$.
The mass fraction in size range dL is $dw = dm/M_T$.

Substitutions lead to : $\dfrac{dw}{dL} = \dfrac{L^3 n}{6 n_o (r_T)^4}$

On substituting $n = n_o e^{-L/(r\tau)}$, one obtains:

$$\frac{dw}{dL} = \frac{e^{-L/(r\tau)} L^3}{6(r\tau)^4}$$

The derivative of the mass fraction distribution is found as follows:

$$\frac{d^2w}{dL^2} = \frac{1}{6(r\tau)^4}\left\{3L^2 e^{-L/(r\tau)} - \frac{L^3}{r\tau} e^{-L/(r\tau)}\right\}$$

The latter expression equals zero if $3 - L/r\tau = 0$. Thus:

$$L_D = 3r\tau \tag{7.14}$$

Nucleation Kinetics

The use of the population density plot as an analytical tool suggests a convenient method to determine nucleation kinetics. A number of experiments at different residence times are carried out in a CMSMPR crystallizer. The residence time is varied by changing the feed rate. The other operational conditions, such as slurry density, feed composition, temperature, and agitation, remain constant. Thus, for example, halving the residence time means doubling the production. It is plausible that the supersaturation will vary while carrying out such a series of experiments. Different supersaturations will result in different growth and nucleation rates and, probably, in different size distributions. By recalculating these size distributions, the kinetic rates can be obtained. Because it is difficult to measure supersaturations, the following development is used.

Rates of nucleation $n_0 r$ and growth r are conventionally written in terms of supersaturation as $n_0 r = k_1 (\Delta C)^b$ and $r = k_2 (\Delta C)^g$. These expressions can be combined to give $n_0 r = k_3 r^i$. $i = b/g$ in which b and g are the kinetic orders of nucleation and growth. The relationship between nucleation and growth may be expressed as:

$$n_0 = k_3 r^{i-1} \tag{7.15}$$

Thus, a plot of $^{10}\log n_0$ vs. $^{10}\log r$ should give a straight line of slope $(i - 1)$. Furthermore, if the order g of the growth process is known, the order of nucleation b can be calculated. Fig. 7.25 gives a plot of $^{10}\log n_0$ vs. $^{10}\log r$. For systems that are believed to follow secondary nucleation mechanisms, the power–law function can be extended to include the slurry density M_T:

$$n_0 = k_4 r^{i-1} M_T{}^j \tag{7.16}$$

Finally, the power–law relationship can include the agitation and flow:

$$n_0 = k_5 r^{i-1} M_T{}^j \varepsilon^k \tag{7.17}$$

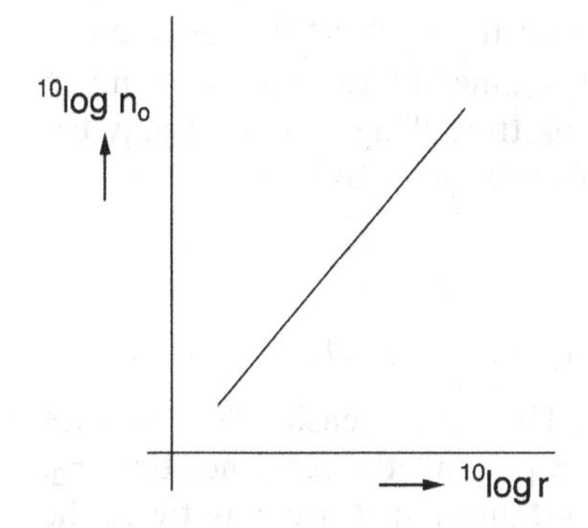

FIG. 7.25 $^{10}\log n_o$ as a function of $^{10}\log r$.

Interpretation of Population Density Plots

Fig. 7.26 gives a population density plot for a CMSMPR crystallizer with fines removal. The residence time of small particles is much smaller than that of larger particles. Fines removal is practiced to obtain large crystals. The net effect of this technique is to force the growth rate to a higher level, producing the same mass on larger average size crystals having less total surface area. Conversely, by plotting $^{10}\log n$ vs. L, it may become apparent that fines removal is occurring.

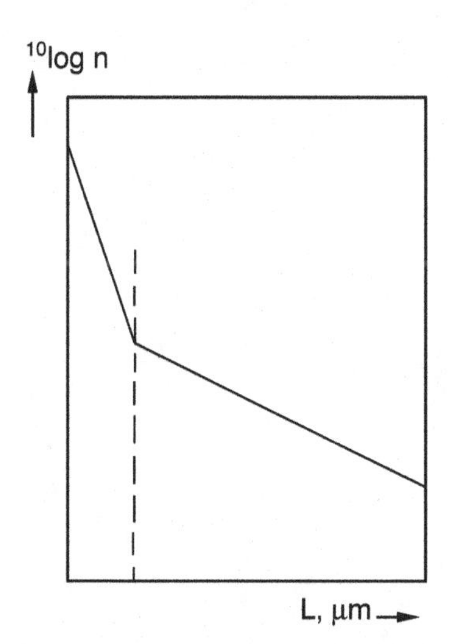

FIG. 7.26 Population density plot in case of fines removal.

Fig. 7.27 gives a population density plot for a CMSMPR crystallizer with classified product removal. The residence time of large particles is much smaller than for small particles. Again, by plotting $^{10}\log n$ vs. L, it may become apparent that there is product classification or abrasion.

Applications

Relationship Between Slurry Density and Dominant Size

The crystallizer is a CMSMPR crystallizer. The slurry density $M_{T,1}$ is twice the slurry density $M_{T,2}$. The residence time τ_1 equals the residence time τ_2. This means that the production of the crystallizer in Case 1 is twice the production in Case 2. The feed composition, temperature, and agitation are kept constant. It is required to calculate the ratio of $L_{D,1}$ and $L_{D,2}$ as a function of i and j:

$$\frac{L_{D,1}}{L_{D,2}} = \frac{3r_1\tau}{3r_2\tau} = \frac{r_1}{r_2}$$

$$\frac{M_{T,1}}{M_{T,2}} = \frac{6\rho_s k_v n_{o,1}(r_1\tau)^4}{6\rho_s k_v n_{o,2}(r_2\tau)^4} = \frac{n_{o,1} \cdot r_1^4}{n_{o,2} \cdot r_2^4} = 2$$

$$\frac{n_{o,1}}{n_{o,2}} = \frac{k_4 r_1^{i-1} M_{T,1}^{\,j}}{k_4 r_2^{i-1} M_{T,2}^{\,j}} = \left(\frac{r_1}{r_2}\right)^{i-1} 2^j$$

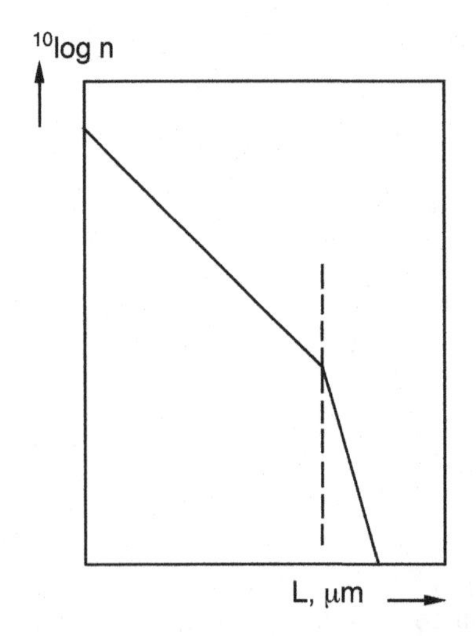

FIG. 7.27 Population density plot in case of classified product removal.

Substitutions lead to : $\dfrac{L_{D,1}}{L_{D,2}} = 2^{(1-j)/(3+i)}$

L_D is independent of M_T if $j = 1$.

RELATIONSHIP BETWEEN RESIDENCE TIME AND DOMINANT SIZE

The crystallizer is a CMSMPR crystallizer. The residence time τ_1 is twice the residence time τ_2. The slurry density $M_{T,1}$ equals the slurry density $M_{T,2}$. This means that the production of Case 2 is twice the production of Case 1. The feed composition, temperature, and agitation are kept constant. It is required to calculate the ratio of the dominant sizes as a function of i and j:

$$\frac{L_{D,1}}{L_{D,2}} = \frac{3r_1\tau_1}{3r_2\tau_2} = 2\frac{r_1}{r_2}$$

$$6\rho_s k_v n_{o,1}(r_1\tau_1)^4 = 6\rho_s k_v n_{o,2}(r_2\tau_2)^4$$

$$n_{o,1} = k_4 r_1^{i-1} M_T{}^j$$

$$n_{o,2} = k_4 r_2^{i-1} M_T{}^j$$

Substitutions lead to : $\dfrac{r_1}{r_2} = \left(\dfrac{\tau_1}{\tau_2}\right)^{-4/(i+3)}$

It follows : $\dfrac{L_{D,1}}{L_{D,2}} = 2^{(i-1)/(i+3)}$

L_D is independent of τ if $i = 1$.

LIST OF SYMBOLS

A	Cryoscopic constant
a	Thermal diffusivity $\{\lambda_l/(c_l\rho_l)$ or $\lambda_s/(c_s\rho_s)\}$ $[\mathrm{m^2 \cdot sec^{-1}}]$
B	Cryoscopic constant
Bi	Biot number $(\alpha_k R/\lambda_s)$
b	Exponent, nucleation order
C	Concentration in a reactor $[\mathrm{gmol \cdot l^{-1}}]$
C_0	Initial concentration in a reactor $[\mathrm{gmol \cdot l^{-1}}]$
C_{bulk}	Impurity concentration in the liquid phase $[\mathrm{gmol \cdot l^{-1}}]$
C_{int}	Impurity concentration at the interface $[\mathrm{gmol \cdot l^{-1}}]$
C_s	Impurity concentration in the solid phase $[\mathrm{gmol \cdot l^{-1}}]$
ΔC	Supersaturation $[\mathrm{gmol \cdot l^{-1}}]$
c	Concentration $[\mathrm{kg \cdot m^{-3}}]$

c_l	Liquid specific heat [$J \cdot kg^{-1} \cdot K^{-1}$]
c_s	Solid specific heat [$J \cdot kg^{-1} \cdot K^{-1}$]
D	Diffusion coefficient [$m^2 \cdot sec^{-1}$]
	Tube diameter [m]
d_m	Diffusion boundary layer thickness [m]
d_{th}	Thermal boundary layer thickness [m]
d_w	Tube wall thickness [m]
Fo	Fourier number (at/R^2)
Gr	Grashof number $\{L^3 g \Delta\rho_l/\nu_l^2 \rho_l\}$
g	Acceleration due to gravity [$m \cdot sec^{-2}$]
	Exponent, growth order
i	Heat of fusion [$J \cdot kg^{-1}$ or $J \cdot kmol^{-1}$]
	Exponent, $i = b/g$
j	Exponent, slurry density order
k	Heat transfer coefficient [$W \cdot m^{-2} \cdot K^{-1}$]
	Exponent, agitation order
k_1	Kinetic constant for nucleation
k_2	Kinetic constant for growth
k_3	k_1/k_2^i
k_4	Coefficient
k_5	Coefficient
k_v	Shape factor
L	Length [m or μm]
L_{av}	Average particle size [μm]
L_D	Dominant particle size [μm or m]
ΔL	Particle size range [μm]
M_T	Slurry density [$g \cdot l^{-1}$]
m	Mass [$g \cdot l^{-1}$]
	Number of gmol
m_1	Mass of one crystal [g]
N or $N(L)$	Cumulative number of crystals per liter of slurry up to size L [l^{-1}]
Nu	Nusselt number ($\alpha_c D/\lambda_l$)
n	Population density [$\mu m^{-1} \cdot l^{-1}$]
n_o	Nuclei population density [$\mu m^{-1} \cdot l^{-1}$]
$\bar{n}$	Average population density in size interval [$\mu m^{-1} \cdot l^{-1}$]
Ph	Phase Transfer number [$i/\{c_s(T_o - T_c)\}$]
Pr	Prandtl number ($\mu c_l/\lambda_l$)
Q	Heat flow [$W \cdot m^{-1}$ or W]
R	Inner heat exchanger tube radius [m]
	Universal gas constant [$J \cdot kmol^{-1} \cdot K^{-1}$]
R_1	Hollow cylinder inner radius [m]

R_2	Hollow cylinder outer radius [m]
	Outer heat exchanger tube radius [m]
R_m	Logarithmic mean radius [m]
Re	Reynolds number (vD/ν)
r	Cylinder radius (variable) [m]
	Growth rate [$\mu m \cdot min^{-1}$ or $m \cdot sec^{-1}$]
T	Temperature [°C or K]
ΔT	Temperature difference [K]
T_1	Inner cylinder temperature [°C]
T_2	Outer cylinder temperature [°C]
T_{act}	Actual temperature [°C]
T_c	Cooling medium temperature [°C]
T_{cryst}	Crystal temperature at suspension crystallization [°C]
T_E	Temperature of the eutectic point [°C or K]
T_{eq}	Equilibrium temperature [°C]
T_o	Melting point [°C]
t	Time [sec]
U	Overall heat transfer coefficient [$W \cdot m^{-2} \cdot K^{-1}$]
V	Reactor or crystallizer volume [l]
v	Film flow rate [$m \cdot sec^{-1}$]
w	Mass distribution
X	Kilomole fraction
	Weight fraction
	Length coordinate [m]
	Variable
α_c	Heat transfer coefficient (medium-side) [$W \cdot m^{-2} \cdot K^{-1}$]
α_k	Heat transfer coefficient (medium-side, includes metal wall resistance) [$W \cdot m^{-2} \cdot K^{-1}$]
α_o	Heat transfer coefficient (process-side) [$W \cdot m^{-2} \cdot K^{-1}$]
δ	Film thickness [m]
ε	Specific stirrer energy input [$W \cdot kg^{-1}$]
λ_l	Liquid thermal conductivity [$W \cdot m^{-1} \cdot K^{-1}$]
λ_s	Solid thermal conductivity [$W \cdot m^{-1} \cdot K^{-1}$]
λ_w	Metal wall thermal conductivity [$W \cdot m^{-1} \cdot K^{-1}$]
μ	Dynamic viscosity [$N \cdot sec \cdot m^{-2}$]
μ_w	Dynamic viscosity at the wall [$N \cdot sec \cdot m^{-2}$]
ν	Kinematic viscosity [$m^2 \cdot sec^{-1}$]
ρ_l	Liquid specific mass [$kg \cdot m^{-3}$]
ρ_s	Solid specific mass [$kg \cdot m^{-3}$ or $g \cdot \mu m^{-3}$]
$\Delta\rho_l$	Liquid specific mass difference [$kg \cdot m^{-3}$]
τ	Residence time [min]
ϕ_v	Volumetric flow [$l \cdot min^{-1}$]

Indices

1	Number
2	Number
3	Number
4	Number
8	Number of C atoms
A	Component A
B	Component B
C	Point in the phase diagram
D	Point on the liquidus
E	Eutectic
F	Point in the phase diagram
L	Liquid
P	Polymorphic
x	Number of atoms in a molecule
y	Number of atoms in a molecule

APPENDIX A. Material Balances (G)

A.1. First Stage

In		Out			
Feed		Draining	Sweating	Melting	Sum
Product	19,356.7	2334.9	2517.1	14,334.6	19,186.6
Impurities	2841.3 +	1474.1 +	791.9 +	675.4 +	2941.4 +
Total	22,198.0	3809.0	3309.0	15,010.0	22,128.0
Assay [%]	87.2	61.3	76.1	95.5	

A.2. Second Stage

In		Out			
Feed		Draining	Sweating	Melting	Sum
Product	21,837.8	4957.3	2276.8	14,582.9	21,817.0
Impurities	1125.2 +	740.7 +	146.8 +	192.1 +	1079.6 +
Total	22,963.0	5698.0	2423.6	14,775.0	22,896.6
Assay [%]	95.1	87.0	93.9	98.7	

A.3. Third Stage

In		Out			
Feed		Draining	Sweating	Melting	Sum
Product	23,246.9	9246.1	2690.6	11,288.5	23,225.2
Impurities	330.1 +	159.9 +	73.1 +	102.5 +	335.5 +
Total	23,577.0	9406.0	2763.7	11,391.0	23,560.7
Assay [%]	98.6	98.3	97.4	99.1	

REFERENCES

Arkenbout, G. F. (1995). *Melt Crystallization Technology*. Lancaster, PA, USA: Technomic.

Beek, W. J., Muttzall, K. M. K., van Heuven, J. W. (1999). *Transport Phenomena*. London: Wiley.

Groot Wassink, J. (1976). *Process Analysis and Design of Drum Flakers*. The Hague, The Netherlands: Stichting Nederlandse Apparaten voor de Procesindustrie.

Jančić, S. J. (1989). Fractional crystallization. In: Nyvlt, J., Záček, S., eds., *Industrial Crystallization*. Amsterdam, The Netherlands: Elsevier, pp. 57–70.

Larson, M. A., Randolph, A. D. (1969). Size distribution analysis in continuous crystallization. In: Palermo, J. A., Larson, M. A., eds., *Crystallization from Solutions and Melts*. New York: American Institute of Chemical Engineers, pp. 1–13.

Matsuoka, M., Fukushima, H. (1986). Determination of solid liquid equilibrium. *Bunri Gijutsu (Sep. Process Eng.)*. 16:4. In Japanese.

McKay, D. L., Dale, G. H., Tabler, D. C. (1966). *Para*-xylene via fractional crystallization. *Chem. Eng. Prog.* 62:104.

Muir, R.F. (1982). Predicting solid–liquid equilibrium data from vapor–liquid data. *Chem. Eng.* 89:89.

Randolph, A. D., Larson, M. A. (1962). Transient and steady state distributions in continuous mixed suspension crystallizers. *Am. Inst. Chem. Eng. J.* 8:639.

Randolph, A. D., Larson, M. A. (1988). *Theory of Particulate Processes*. New York: Academic Press.

Rittner, S., Steiner, R. (1985). The melt crystallization of organic materials and its large-scale application. *Chemie-Ingenieur-Technik* 57:91. In German.

Sloan, G. J., McGhie, A. R. (1988). *Techniques of Melt Crystallization*. London: Wiley.

Takegami, K., Meyer, D. (1996). The development of crystallisation for *p*-xylene purification. *Int. J. Hydrocarb. Eng.* 1:29.

Thermodynamics Research Center (TRC) (1996). *Thermodynamic Tables—Hydrocarbons*. College Station, TX: The Texas A&M University System.

VDI (1993). *VDI Heat Atlas*. Düsseldorf, Germany: VDI Verlag.

Wintermantel, K. (1986). The separation efficiency at layer crystallization from melts and solutions—a consistent representation. *Chemie-Ingenieur-Technik* 1493:86. In German.

Wintermantel, K., Wellinghof, G. (1991). Melt crystallization—theoretical conditions and technical limits. *Chemie-Ingenieur-Technik* 63:881. In German.

8

Fluid Bed Granulation

8.1. INTRODUCTION

Fluid bed granulation is a modern alternative for prilling. It is applied to the granulation of the fertilizers urea and ammonium nitrate (AN) mainly. It comprises the upward atomization of a liquid feed into a bed of moving particles. The droplets wet the particles and crystallization takes place.

Hydro Agri's fluid bed granulation technology is reviewed in Sec. 8.2. Urea product characteristics are given in Sec. 8.3, and the properties of granulated urea are compared to the properties of prilled urea. Physical data of AN are reviewed in Sec. 8.4, and the properties of granulated AN are also compared to the properties of prilled AN. Ammonium nitrate is applied both as a fertilizer and as an explosive. Several aspects of its explosive nature are discussed in this section as well.

Fluidization theory is introduced in Sec. 8.5. For example, the concept of minimum fluidization velocity is discussed in this section. In Sec. 8.6, the pressure drop of gas distributors is dealt with.

Hydro Agri's fluid bed granulation technology comprises liquid feed atomization by means of pneumatic nozzles. This type of atomization is discussed in Sec. 8.7. The heat balance of urea fluid bed granulation is reviewed in Sec. 8.8. Fluid bed cooling of granulated urea and AN is treated in Sec. 8.9.

It is assumed that at any position in the bed, the temperature of the product leaving a mixing cell equals the temperature of the leaving air.

Toyo Engineering Corporation (TEC) offers a spouted bed granulation technology for urea and AN. It resembles Hydro Agri's method. One difference is the upward atomization of liquid feed by means of single-fluid nozzles in spout pipes. TEC's technology is discussed in Sec. 8.10.

To accomplish gas/solid separation, scrubbers are used. Scrubbers form a class of devices in which a liquid (usually water) is used to assist or accomplish the collection of dusts or mists. These devices are dealt with in Sec. 8.11.

8.2. A TYPICAL HYDRO AGRI GRANULATION PLANT

A typical Hydro Agri granulation plant for urea is depicted in Fig. 8.1. A typical plant capacity is 2000 t per day. The feed solution, normally containing 96.5% urea and 3.5% water by weight and having a temperature of 135°C, is atomized upwards into a bed of moving particles by means of two-fluid nozzles. Urea's melting point is 132.6°C and the bed temperature is somewhat lower. The droplets wet the particles and drying and crystallization take place

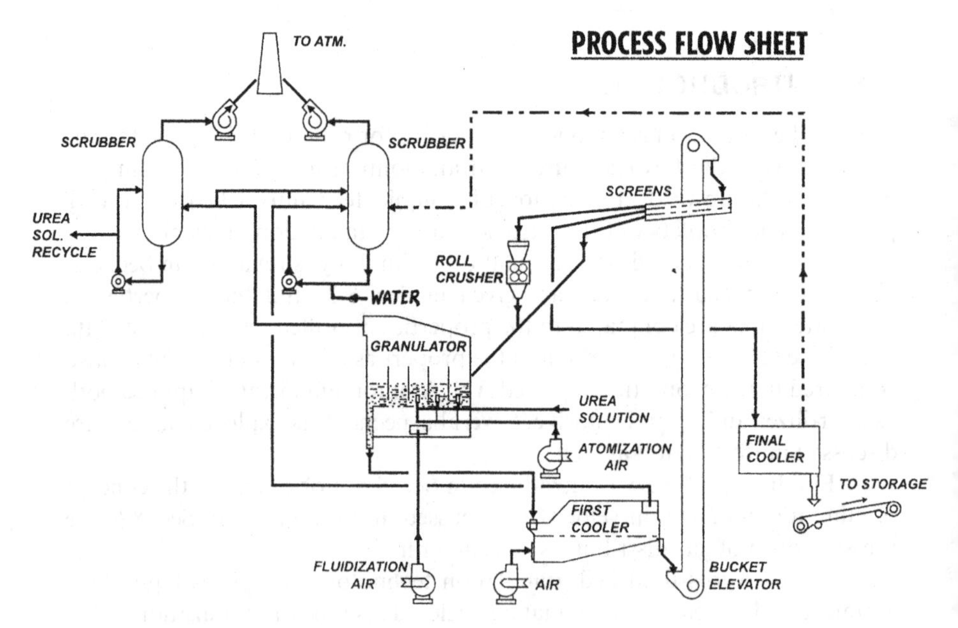

FIG. 8.1 A typical Hydro Agri granulation plant for urea. (Courtesy of Hydro Fertilizer Technology, Brussels, Belgium.)

subsequently. The atomization of the feed solution is accomplished by low-pressure air. The atomization air is at a temperature approximately equal to urea's melting point to prevent crystallization in the spray nozzle.

The urea particles are kept in a fluidized state by fluidization air. The fluidization air is at ambient temperature. It is not necessary to heat the fluidization air because the heat of crystallization of urea is sufficient to heat up the fluidization air, to evaporate the water, and to heat up the recycle. See Sec. 8.8 for a heat balance of the urea granulation. The fluidization air is evenly distributed by a perforated distributor plate having a free area of 5–10%. The pressure loss of the distributor plate is a fraction of the pressure loss of the fluidized layer (see Sec. 8.6). The fluid bed granulator is of the rectangular type; that is, the length is 3 to 4 times larger than the width. The fluid bed granulator consists of a number of compartments. The compartments are separated from each other by means of baffles. There is vigorous mixing in each compartment. The material travels from one compartment to the next by underflow.

Per kilogram of urea produced by the plant, approximately 0.5 kg of fine and crushed urea particles is fed to the granulator. These particles grow as they pass through the granulator. The number of particles removed from the granulator per hour equals the number of particles fed to the granulator. Thus the nip setting of the crusher controls the average particle size of the urea leaving the granulator. A typical weight average particle size is 3.2 mm.

Like the fluidization air, the urea melt is fed in cross-flow over the full fluid bed granulator area. The capacity of one spray nozzle is approximately $0.5\ t \cdot hr^{-1}$. The fluidized layer height is about 1 m and the residence time is a couple of minutes. The residence time is expressed as the ratio of the granulator product holdup in kilograms and the mass flow from the granulator in kilograms per minute.

Scale-up of the granulator is achieved by increasing the area and the number of two-fluid nozzles.

Process control of the fluid bed granulator is as follows. The flow rates of urea melt, atomization air, and fluidization air are fixed. The concentration of the urea melt is adjusted automatically or manually to obtain the desired bed temperature. The average particle size is controlled by the nip setting of the crusher.

Formaldehyde is injected into the feed solution in order to improve granulation and the caking characteristics. Formaldehyde retards the crystallization of urea; thus urea does not crystallize instantaneously after having left the spray nozzle and formation of dust is thereby largely prevented. Moreover, the crushing strength increases due to more regular crystallization. In a warehouse, urea is prone to caking because of its hygroscopicity. It absorbs water from the air when the relative humidity is high (see Sec. 8.3).

This water dissolves urea. Water evaporates again when the relative humidity subsequently drops. The dissolved urea recrystallizes as bridges between the particles. The effect of formaldehyde is to render these bridges relatively weak.

The air leaving the granulator contains some rather coarse urea dust which is caught in a scrubber of the impingement type (see Sec. 8.11). More than 99.5% by weight is thus recovered. This recovered urea is recycled to the concentration section of the plant as a 45% by weight solution. Dust outlet concentrations of less than 30 $mg \cdot nm^{-3}$ in the stack can be achieved, roughly corresponding to a urea loss of less than 0.1 kg per ton of urea produced.

Granular urea flows out of the granulator to a fluid bed cooler, in which the product is cooled by means of ambient air. Now the granules are strong enough to survive mechanical handling. The cooled granules are separated into three fractions by means of screens. Oversize and undersize products are recycled to the granulator. The coarse fraction is crushed after it has left the screens.

The on-spec product from the screens is cooled by means of chilled air in a second fluid bed cooler. The cooling of urea to a constant and sufficiently low storage temperature is one of the most significant parameters to prevent caking. Typically, urea is cooled to 40°C.

Any air from the second fluid bed cooler also flows to a scrubber. Process water is added to the latter scrubber. A diluted urea solution flows to the scrubber of the granulator, and here the urea solution is recycled.

The process for fluid bed granulation of ammonium nitrate (AN) or calcium ammonium nitrate (CAN) is similar to the process for fluid bed granulation of urea. When CAN is produced, the plant is equipped with an additional unit wherein a 95% by weight AN solution is mixed with finely ground dolomite and with the diluted scrubber solution. Dolomite is the double salt $CaMg(CO_3)_2$. This material occurs naturally; for example, the mountains in Northern Italy called the Dolomites consist of dolomite mainly. The mineral is obtained by means of mining and is milled for use in fertilizers. The mixture is then reconcentrated before being fed to the granulator. The water level of the feed is 2.5% by weight. The nitrogen level of the finished CAN granules is 27% by weight. The nitrogen level of finished AN granules is 33.5% by weight. CAN granules have better product safety characteristics than AN granules because they are more diluted (see Sec. 8.4).

The feed has a temperature of 150°C and the bed temperature is somewhat lower. The melting point of pure AN is 170°C. The heat of crystallization of AN is much lower than the heat of crystallization of urea. Thus it is necessary to heat the fluidization air to 90°C. The steam flow to the air heater is used to control the bed temperature.

The last part of the fluid bed granulator is a fluid bed cooler. The granules leaving the cooling section are strong enough to survive mechanical handling. However, they are still hot. The product is screened and oversize and undersize are recycled. The on-spec product is cooled in two fluid bed coolers in series. Conditioned air is used in the second cooler because AN is hygroscopic.

8.3. UREA

Urea, $CO(NH_2)_2$, is an important chemical product that is used as a fertilizer. The worldwide production in 1999 was approximately $90 \cdot 10^6$ tpa. Urea was synthesized by Wöhler in 1828 from ammonia and cyanic acid in an aqueous solution. He thereby proved that there is no fundamental difference between organic and inorganic chemistry. Up till then, it was believed that organic compounds could not be synthesized in a laboratory because a "vital force" would be required. Urea plays an important role in many biological processes, e.g., in the decomposition of proteins. The human body produces 20–30 g of urea per day.

Urea (46.2% nitrogen by weight) is an even more concentrated source of nitrogen than ammonium nitrate and thus has a lower cost of distribution of nitrogen. One disadvantage vis-à-vis ammonium nitrate is the possibility of substantial loss of gaseous ammonia into the atmosphere from urea hydrolysis if it is not washed into the soil by rain promptly. This effect can produce 10–15% nitrogen loss. Furthermore, the maximum crop yield at the highest urea application can be reduced by some 5% due to a small amount of poisoning by high concentration of ammonia. Urea's melting point is 132.6°C and its heat of crystallization is $224.5 \text{ kJ} \cdot \text{kg}^{-1}$. Urea is very soluble in water; for example, a saturated aqueous solution contains 51.8% urea by weight at 20°C. Urea is hygroscopic; for example, urea absorbs water from the atmosphere at 20°C when the relative humidity of this air exceeds 74%.

First, the prilled product will be discussed. Prilling as a unit operation is treated in Chapter 5. Granules made by fluid bed granulation will be discussed next. In a prilling plant, the urea solution is concentrated to minimum 99.8% urea by weight. The feed water content cannot be used as an independent process variable because the residence time is too short to accomplish drying down to the final moisture specification. Thus the prilling tower feed has to be practically dry. The urea solution is pumped to the top of a 50–60 m cylindrical concrete tower where it is fed into a spinning bucket containing many small holes. The very fine dust that is formed and exits at the top of the tower with the air requires very efficient scrubbing to avoid environmental problems. The weight average particle size is maximum 2 mm because the operation is carried out in the gravity field of the earth. This is explained as follows. For practical reasons, tower heights are maximum 60 m. Prills larger than 2 mm, on falling through the tower, do not get enough time to crystallize and cool completely. Their terminal velocity is in the range 8–10 m · sec^{-1} (see Table 5.3) and 6–8 sec is not sufficient. The bulk density of prilled urea is typically 740 kg · m^{-3}.

The crushing strength of the prills is smaller than the crushing strength of granules made by fluid bed granulation. Typically, the crushing strength of a 2-mm prill is 0.6 kg, while the crushing strength of a 2.5-mm granule is typically 3 kg or slightly higher. Because of this, granules made by fluid bed granulation are, at present, preferred over prills. For accurate dosing, the generation of powder is undesirable. At prilling, formaldehyde added upstream also reduces dust emission and caking tendencies in bulk storage. The effect of formaldehyde in a granulation process was discussed in Sec. 8.2.

Now the attention will be directed to fluid bed granulation. Fluid bed granulation starts with melts containing 3.5% water by weight. Compared to prilling, there is less tendency to generate fine particles.

In a fluid bed granulation process, less air is needed per kilogram of urea than in a prilling process. In a prilling process, the heat required to cool the melt and to crystallize urea is carried away by the air only. Furthermore, in a prilling process, the air can be heated up by 20 K only. In granulation, a substantial part of the heat to cool the melt and to crystallize the urea is absorbed by water evaporation. In addition, the fluidization air is heated up from 20°C to 110°C (see Sec. 8.8 for a heat balance of urea granulation). The weight average particle size of fluid bed granules can be selected from the range 2–8 mm. This is because the granules grow from single-seed particles. The bulk density of granulated urea is typically 780 kg · m^{-3}.

8.4. AMMONIUM NITRATE

Ammonium nitrate, NH_4NO_3 or AN, is a major chemical product. It is used primarily as a nitrogen fertilizer. It is also the principal component of most industrial explosives and nonmilitary blasting compositions. AN does not occur in nature because it is very soluble in water; for example, a saturated aqueous solution contains at 20°C 65.2 g of the compound per 100 g of solution. AN is very hygroscopic; for example, the solid salt picks up water from the air at 20°C when the relative humidity exceeds 65%. The hygroscopic nature complicates its usage. However, coating techniques have been developed.

The melting point of AN is 170°C and its heat of crystallization is 77.5 kJ · kg^{-1}. Solid AN occurs in five different crystalline forms detectable by time–temperature cooling curves. One transition is at 32.3°C. This transition is particularly significant for the storage of fertilizers containing AN: on passing through the transition repeatedly, the granules lose strength and finally disintegrate because of the differing densities. Normally, a stabilizer is added to minimize this effect.

The AN molecule contains both oxidizing agent and reductant ready and in close proximity to react. However, the material is considered a stable salt, although it loses ammonia and becomes slightly acidic on storage. When the salt is heated to temperatures from 200°C to 230°C, exothermic decomposition occurs. Above 230°C, exothermic elimination of N_2 and NO_2 begins. The final violent exothermic reaction occurs with great rapidity when ammonium nitrate detonates.

The reaction equation of the latter event is:

$$2NH_4NO_3 \rightarrow 2N_2 + 4H_2O + O_2 \quad \Delta H = -118.5 \text{ kJ} \cdot \text{gmol}^{-1}$$

Ammonium nitrate is sensitized by, for example, cobalt, chloride, and oil, so these and related substances should be excluded from manufacturing processes. Accidental detonations of large amounts of fertilizer material containing AN have occurred in the past. In some cases, these detonations took place when the temperature of AN was raised by the combustion of other materials. Often, inert materials, such as dolomite, are added to AN for fertilizer use. Inert materials can phlegmatize AN. A fertilizer containing maximum 80% AN by weight, maximum 0.4% combustible material by weight, and minimum 18% magnesium or calcium carbonate by weight is, in principle, nondetonatable. The detonatibility of fertilizers containing AN can be checked by means of tests [see, e.g., E.E.C. directive 80/876 (1980)]. In dealing with AN as an explosive in this section, the detonation phenomenon will be discussed in more detail.

8.4.1. Ammonium Nitrate as a Fertilizer

First, the prilled product will be discussed. Prilling as a unit operation is treated in Chapter 5. Granules made by fluid bed granulation will be discussed next.

Prilling starts from melts containing approximately 0.5% water by weight. As in the prilling of urea, the feed has to be practically dry. Round or rectangular spray towers up to 60 m tall are used. The melt is pumped to the top of the tower where it is fed into a spinning bucket containing many small holes. As in the prilling of urea, there is an inherent tendency to form very fine dust requiring very efficient scrubbing to avoid environmental problems. For the same reason as explained for the prilling of urea, the weight average particle size is maximum 2 mm. The bulk density of prilled AN is typically 860 $kg \cdot m^{-3}$. The moisture content of the prilled product is 0.2–0.3% water by weight. The crushing strength of the prills is smaller than the crushing strength of granules made by fluid bed granulation. Because of this, granules made by fluid bed granulation are, at present, preferred over prills. For accurate dosing, the formation of powder is undesirable.

Now attention will be directed to fluid bed granulation. Compared to prilling, there is less tendency to generate fine particles. Fluid bed granulation starts from melts containing 2.5% water by weight. Fluid bed granulation of AN needs less air than prilling per kilogram of AN. Basically, the reasons have been given during the discussion on urea in Sec. 8.3. Because of the harder particles and the smaller amount of air, there are less pollution problems during fluid bed granulation than at prilling. Because the granules grow from single-seed particles, the weight average particle size of fluid bed granules can be selected from the range 2–8 mm. A normal weight average diameter is 3.3 mm. The bulk density of AN granules is typically 1000 $kg \cdot m^{-3}$. These granules contain 33.5% nitrogen by weight.

The bulk density of CAN granules is typically 1070 $kg \cdot m^{-3}$. CAN granules contain 27% nitrogen by weight. Calcium ammonium nitrate granules are diluted with milled dolomite. The moisture content of the fluid bed granulate is typically 0.3% water by weight. The crushing strength of an AN granule having a diameter of 3.15 mm is typically 3.5 kg.

Magnesium nitrate, $Mg(NO_3)_2$, is added to the melt to enhance AN's resistance to thermal cycling and to improve the caking characteristics. The latter characteristics are improved because magnesium nitrate absorbs moisture. A further step to prevent caking is coating the granules in a rotary drum with effective organic surfactants. Thus the coating is added downstream of the fluid bed granulator, while dolomite and magnesium nitrate are added upstream to the fluid bed granulator.

8.4.2. Ammonium Nitrate as an Explosive

The explosives industry requires a porous, low-density AN that can be impregnated with oil. Porous AN can be produced by prilling. An AN melt of relatively low concentration, i.e., 96% by weight, is sprayed into the prilling tower. The prilled product is subsequently dried. The product contains 0.2% water by weight and has a bulk density of 770 $kg \cdot m^{-3}$ and a porosity of 33% by volume.

To aid the understanding, several remarks concerning the detonation phenomenon will be made. A start will be made by defining the explosion phenomenon. An explosion is a physical or chemical phenomenon in which energy is released in a very short time, usually accompanied by formation and vigorous expansion of a very large amount of hot gas. The following categories of explosions may be distinguished:

1. Mechanical explosions. Mechanical explosions are caused by the sudden breaking of a vessel containing gas under pressure.

2. Chemical explosions. This type of explosion is caused by decomposition or very rapid reaction of a product or a mixture.
3. Nuclear explosions. Nuclear explosions are caused by fission or fusion of atomic nuclei.
4. Electrical explosions. This type of explosion is caused by sudden strong electrical currents that volatilize metal wire (exploding wire).

A detonation is a chemical explosion. For a chemical explosion to occur, the reaction must be exothermic, a large amount of gas must be produced by the chemical reaction and evaporation of reaction products, and the reaction must propagate very fast.

A detonation is a chemical explosion that is characterized by a shock wave formed within the decomposing or reacting product and transmitted perpendicularly to the decomposition surface at a very high rate of speed (exceeding the velocity of sound, e.g., several thousands of meters per second).

Primary and secondary explosives are distinguished. Primary explosives (initiator explosives) detonate following weak external stimuli, like percussion, friction, electrical energy, or light energy. Secondary explosives are much less sensitive to shock. However, they can detonate under strong stimulus, such as a shock wave produced by a primary explosive, which may be reinforced by a booster composed of a more sensitive secondary explosive. When an explosive detonates, the shock wave may initiate less-sensitive explosives, cause destruction, or split rocks and soils. The energy released in detonation can be provided by decompositions or oxidation-reduction reactions. A typical example of a decomposition is the detonation of lead azide. Its principal detonation products are lead and nitrogen. Oxidation–reduction reactions are either internal oxidation–reductions, as in the decomposition of AN, or reactions between discrete oxidizers and fuels in heterogeneous mixtures. An example of the latter category is the detonation of a pyrophoric mixture called flash powder. Flash powder is a mixture of potassium chlorate or potassium perchlorate crystals (the oxidizer), aluminum powder, and sulfur powder. The latter two materials are oxidized. AN is a secondary explosive. Secondary explosives are sometimes called high explosives. AN is the cheapest and safest source of readily deliverable oxygen for explosive applications. The extensive use of AN in AN-fuel oil (ANFO) and water-based commercial explosives has largely displaced the nitroglycerin-based dynamites. AN industrial explosives are low cost, safe, versatile in performance and application, and have better storage stability than dynamites. A large number of formulations are available for almost all purposes. Although hygroscopicity is the principal disadvantage of AN, coating techniques have been developed to reduce susceptibility to high humidity.

When a combustible material such as fuel oil is present in stoichiometric proportions (ca. 5.6% by weight), the energy evolved at the detonation reaction increases almost threefold:

$$3NH_4NO_3 + (CH_2)_1 \rightarrow 3N_2 + 7H_2O + CO_2 \quad \Delta H = -343.0\,\mathrm{kJ \cdot gmol^{-1}}$$

Pure AN is stable and insensitive to impact and friction. It is impossible to initiate using conventional blasting caps unless strongly confined and in powder form. When mixed with organic materials such as hydrocarbons or cellulose, AN can be initiated by a powerful high explosive booster. ANFO 94/6 consists of 94% AN by weight coated with an anticaking agent and 6% absorbed fuel oil (FO) by weight. Water-based explosives and water-in-oil emulsion explosives were developed to capitalize on the low cost of AN, to increase the available energy per unit volume beyond that obtainable by ANFOs, to improve ignitability in small diameter charges, and particularly to eliminate the problems associated with the solubility of ANFOs in wet drill holes.

Water-based explosives are also called slurry explosives. Their water content is in the range 10–20% by weight. They are thickened suspensions of oxidizers (AN and other nitrates), fuels, and a sensitizer dispersed in a saturated aqueous salt solution.

Water-in-oil emulsion explosives have been made as typified by a formulation containing 20% water, 12% oil, 2% microspheres, 1% emulsifier, and 65% AN. The percentages are percentages by weight. The microdroplets of an emulsion explosive offer the advantage of intimate contact between fuel and oxidizer and tend to equal or outperform conventional water-based slurries.

8.5. FLUIDIZATION THEORY

The term fluidization was coined to describe a certain mode of contacting granular solids with fluids. To illustrate, consider the solid be a well-rounded silica sand contained in a cylindrical vessel with a porous bottom. The fluid is air. As the air passes upward, through the porous bottom and the bed, measurements can be taken of the bed height and of the pressure drop across the bed as a function of the airflow rate. A profile as shown in Fig. 8.2 will be obtained. When the pressure drop reaches its maximum value, the bed starts to expand. In this condition, the individual sand granules are disengaged from each other and may be moved readily around by the expenditure of much less energy than would be required if the bed was not suspended by the airstream. The mobility of the aerated sand column resembles that of a high-viscosity liquid. This condition is known as the fluidization point and the corre-

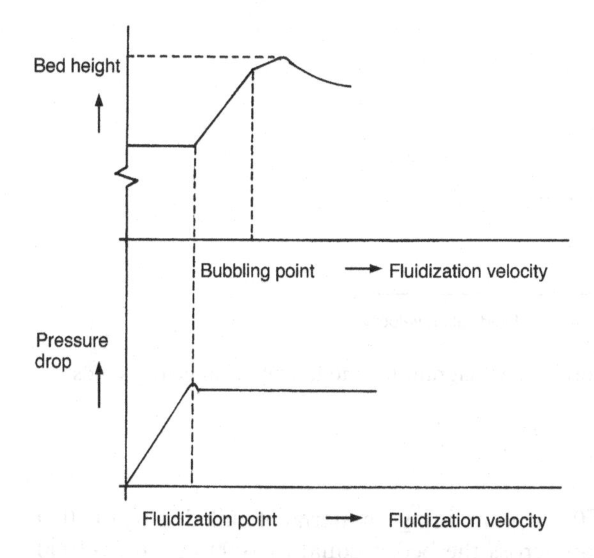

FIG. 8.2 Diagrams for ideally fluidizing solids.

sponding airflow rate is termed the minimum fluidization velocity. With a further increase of the airflow, the point at which the first bubbles appear can be observed. This point locates the bubbling point and the minimum bubbling velocity. A further increase in the flow rate leads to the maximum bed-expansion ratio.

The first large-scale fluidization application was used in the United States in 1940 for cracking oil vapor. Fluid cracking catalysts have a weight average particle size of about 80 μm and a bulk density of about 1100 $kg \cdot m^{-3}$. The description and successful operation of the fluid cracking units precipitated a number of fundamental and applied studies in fluidization fields.

The very efficient heat and mass transfer in a fluidized bed was seen as the principal advantage of this system. Fig. 8.2 can be used to show how granular materials display ideal fluidization characteristics; however, not all materials exhibit this behavior. For example, channeling (preferential airflow) is a frequently found unwanted phenomenon, and Fig. 8.3 shows typical pressure drop/flow data for moderately channeling solids (with flour, particle size ≤ 20 μm, exhibiting the channeling). The design of the gas-inlet device has a profound effect on channeling; thus with porous plates, by means of which gas distribution into a bed tends toward uniformity, channeling tendencies are much reduced when compared with a multiorifice distributor in which the gas is introduced through a relatively small number of geometrically spaced holes.

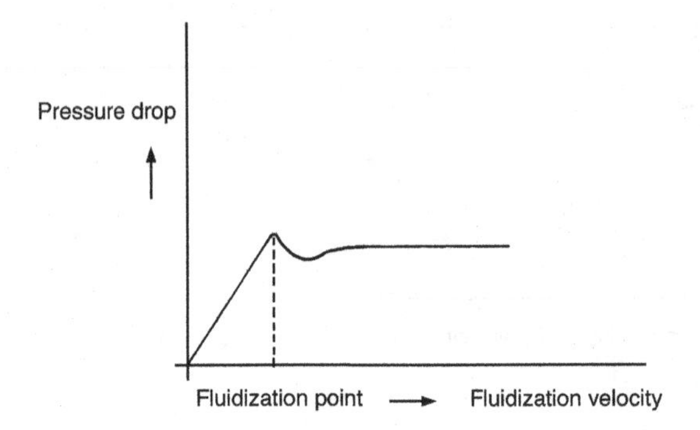

FIG. 8.3 Typical pressure drop/flow diagram for moderately channeling solids.

Example 1

With a bed height of 0.1 m, containing a material of bulk density of 1000 $kg \cdot m^{-3}$, the pressure drop across the bed is equal to $1000 \cdot g \cdot 0.1 \approx 1000$ $N \cdot m^{-2}$ (100 mm wg).

The ability to predict reliably the point of incipient fluidization is of basic importance in virtually all fluidized-process studies and designs. Fluidized reactors and granulators are usually operated at fluid rates that are well in excess of minimum fluidization rates. However, the fluid rate should not exceed the terminal velocity leading to the dilute phase. Leva (1959) recommended for the estimation of the minimum fluidization mass velocity:

$$G_{mf} = 688\,\bar{d}_p^{1.82} \frac{\{\rho_F(\rho_S - \rho_F)\}^{0.94}}{\mu_g^{0.88}} \quad \text{(Imperial units, } \mu_g \text{ in cP)}$$

In SI units:

$$G_{mf} = 0.0093\,\bar{d}_p^{1.82} \frac{\{\rho_F(\rho_S - \rho_F)\}^{0.94}}{\mu_g^{0.88}} \ \text{kg} \cdot \text{m}^{-2} \cdot \text{sec}^{-1} \tag{8.1}$$

This equation is supported by many experimental studies. The correlation is limited to applications where:

$$Re = \frac{G_{mf}\,\bar{d}_p}{\mu_g} < 10$$

If $Re > 10$, G_{mf} must be corrected using Fig. 8.4. G_{mf} must be multiplied by the value found. If the material is vesicular, the solid specific mass must be corrected. The composite diameter is the Sauter mean diameter. The latter diameter is also called the surface-to-volume diameter. The Sauter mean diameter is the diameter of the particle having the same $\overline{d}_p^3/\overline{d}_p^2$ ratio as the entire sample.

For a sample, the Sauter mean diameter is calculated in Table 8.1. It is shown that it is either necessary to count particles or to weigh the screen fractions.

Example 2

A perforated plate is charged with dry urea. The bed height is 1 m and the bulk density is 740 kg · m^{-3}. The specific mass is 1315 kg · m^{-3} at 120°C. Sieve analysis provides the following results:

Size range (mm)	Percent by weight	X,–	d_p (mm)	X/d_p, (mm^{-1})
>4.00	0	0	—	0
3.15–4.00	8	0.08	3.575	0.02238
2.50–3.15	24	0.24	2.825	0.08496
2.00–2.50	35	0.35	2.250	0.15556
1.60–2.00	28	0.28	1.80	0.15556
1.25–1.60	4	0.04	1.425	0.02807
<1.25	1	0.01	1.00	0.01000 +
				0.45653

$$\overline{d}_p = 1/0.45653 = 2.19 \text{ mm}$$

For the applicability of this calculation, the form factor of the particles should remain constant (see also Table 8.1).

It is required to calculate the minimum fluidization velocity based on 2.19 mm.

Ambient air is heated up to 120°C and used for fluidization. The bed temperature is also 120°C. The air leaves the bed at atmospheric pressure.

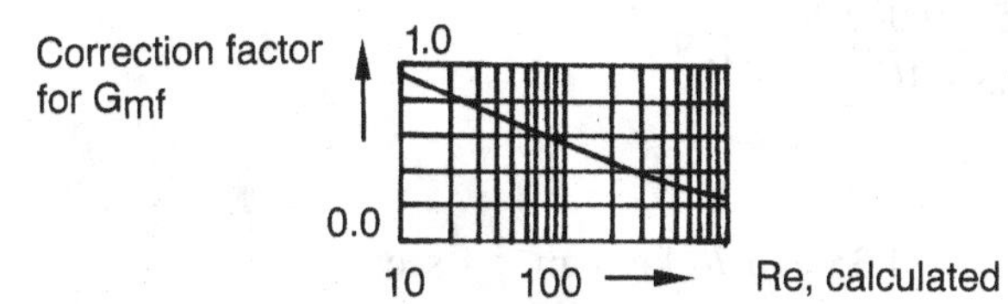

FIG. 8.4 Correction factor for results obtained with Leva's equation if $Re > 10$. (Courtesy of Génie Chimique.)

TABLE 8.1 The Calculation of the Sauter Mean Diameter

Size range (μm)	d_p (μm)	n	$d_p^2 \cdot n$	$d_p^3 \cdot n$	X
0–5	2.5	3	19	47	0.00
5–15	10	18	1800	18,000	0.01
15–25	20	130	52,000	1,040,000	0.03
25–35	30	220	198,000	5,940,000	0.20
35–45	40	105	168,000	6,720,000	0.22
45–55	50	48	120,000	6,000,000	0.20
55–65	60	16	57,600	3,456,000	0.11
65–75	70	11	53,900	3,773,000	0.13
75–85	80	6	38,400+	3,072,000+	0.10+
			689,719	30,019,047	1.00

$\overline{d}_p = \frac{30,019,047}{689,719} = 43.5\ \mu m$

Alternatively:

$$\frac{1}{\overline{d}_p} = \sum \frac{X}{d_p} = \frac{0.01}{10} + \frac{0.03}{20} + \frac{0.20}{30} + \frac{0.22}{40} + \frac{0.20}{50} + \frac{0.11}{60} + \frac{0.13}{70} + \frac{0.10}{80} = 0.0236\ \mu m^{-1}$$

$$\overline{d}_p = 42.4 \mu m$$

The latter expression is valid only if the form factor of the particles is not a function of the particle size; that is, the mass of any particle can be expressed as $k_f \rho_s d_p^3$. Then, for the calculation of $\overline{d}_p$, a screen analysis is sufficient and it is not necessary to count particles.

$$\text{Bed pressure drop}: \Delta p = \rho_b g h = 740 \cdot 9.81 \cdot 1 = 7259\ \text{N} \cdot \text{m}^{-2}$$

$$\text{Inlet air pressure}: 101,300 + 7259 = 108,559\ \text{N} \cdot \text{m}^{-2}$$

$$\rho_F = 1.30 \cdot \frac{273}{273+120} \cdot \frac{108,559}{101,300} = 0.97\ \text{kg} \cdot \text{m}^{-3}$$

$$G_{mf} = 0.0093 \left(2.19 \cdot 10^{-3}\right)^{1.82} \cdot \frac{\{0.97(1315 - 0.97)\}^{0.94}}{\left(2.3 \cdot 10^{-5}\right)^{0.88}}$$

$$= 1.35\ \text{kg} \cdot \text{m}^{-2} \cdot \text{sec}^{-1}$$

$$Re = \frac{G_{mf}\overline{d}_p}{\mu_g} = \frac{1.35 \cdot 0.00219}{2.3 \cdot 10^{-5}} = 129$$

Correction factor 0.55

$$\text{Corrected } G_{mf} = 0.55 \cdot 1.35 = 0.74\ \text{kg} \cdot \text{m}^{-2} \cdot \text{sec}^{-1}$$

$$v_{mf} = \frac{0.74}{0.97} = 0.76\ \text{m} \cdot \text{sec}^{-1}$$

Calculation of the minimum fluidization velocity below the perforated plate

Assume a distributor pressure drop of 20% of the fluidization pressure drop: $0.2 \cdot 7259 = 1452\ \mathrm{N \cdot m^{-2}}$.

$$\rho_g = 1.30 \cdot \frac{273}{273 + 120} \cdot \frac{(101{,}300 + 1452 + 7259)}{101{,}300} = 0.98\ \mathrm{kg \cdot m^{-3}}$$

$$v_{mf}' = 0.76 \cdot \frac{0.97}{0.98} = 0.75\ \mathrm{m \cdot sec^{-1}}$$

A similar empirical relationship exists for the bubbling point velocity. The fluidization velocity actually chosen is usually many times greater than the minimum fluidization velocity. Table 8.2 (Perry and Green, 1997) provides an estimate of the maximum superficial air velocities obtained through vibrating conveyor screens. Using this table, we find that the maximum superficial air velocity for the material of the example is about 10 times the minimum fluidization velocity.

The fluidization velocity is often chosen after consideration of the carryover of fines by calculating the terminal velocity of the fines.

Example 3

Calculation of the terminal velocity of urea particles having a diameter of 1.25 mm in air.

The air is at atmospheric pressure and has a temperature of 120°C. Maximum 1% by weight of the product of Example 2 will be carried over if

TABLE 8.2 Estimate of Maximum Superficial Air Velocities Through Vibrator–Conveyor Screens ($\mathrm{m \cdot sec^{-1}}$)

		Specific gravity	
Mesh size	Microns	2.0	1.0
200	74	0.22	0.13
100	149	0.69	0.38
50	297	1.4	0.89
30	595	2.6	1.8
20	841	3.2	2.5
10	2000	6.9	4.6
5	4000	11.4	7.9

Column 2 gives the corresponding hole sizes in μm. The mesh size reflects the U.S. number according to ASTM_E_11_61.
Source: Columns 1, 3, and 4: Courtesy of Carrier Division, Rexmond Inc.

the air velocity in the disengagement section equals this terminal velocity. The approach of Sec. 5.4 is followed.

Assume $v_p = 6.3\ \text{m} \cdot \text{sec}^{-1}$ (Table 5.1).

$$\rho_g = 1.30 \cdot \frac{273}{273 + 120} = 0.90\ \text{kg} \cdot \text{m}^{-3}$$

$$\mu_g = 2.3 \cdot 10^{-5}\ \text{N} \cdot \text{sec} \cdot \text{m}^{-2}$$

$$Re = \frac{0.90 \cdot 6.3 \cdot 1.25 \cdot 10^{-3}}{2.3 \cdot 10^{-5}} = 308.2$$

$c_w = 0.6$ (Fig. 5.7).

Equation (5.3):

$$\frac{\pi}{6} d_p^3 (\rho_s - \rho_g) g \stackrel{?}{=} c_w \frac{\pi}{4} d_p^2 \frac{1}{2} \rho_g v_p^2$$

$$\frac{\pi}{6} (1.25 \cdot 10^{-3})^3 (1315 - 0.90) 9.81 = 1.32 \cdot 10^{-5}$$

$$0.6 \cdot \frac{\pi}{4} (1.25 \cdot 10^{-3})^2 \frac{1}{2} \cdot 0.90 \cdot 6.3^2 = 1.32 \cdot 10^{-5}$$

$$v_p = 6.3\ \text{m} \cdot \text{sec}^{-1}$$

Calculation of the maximum fluidization velocity below the perforated plate

The air temperature is 120°C. Assume a distributor pressure drop of 50% of the fluidization pressure drop: $0.5 \cdot 7259 = 3630\ \text{N} \cdot \text{m}^{-2}$.

$$\rho_g = 1.30 \cdot \frac{273}{273 + 120} \cdot \frac{101{,}300 + 3630 + 7259}{101{,}300} = 1.00\ \text{kg} \cdot \text{m}^{-3}$$

$$v_{\text{maxf}} = 6.3 \cdot \frac{0.90}{1.00} = 5.7\ \text{m} \cdot \text{sec}^{-1}$$

8.6. GAS DISTRIBUTORS

The Hydro Agri fluid bed granulator has a perforated plate to distribute the fluidization air evenly. Thanks to angled holes in the distributor plate, horizontal aerodynamic forces exert a thrust on the heavier granules and on small lumps and ensure an efficient horizontal transport of the material. The holes are nose-shaped (see Figs. 8.5 and 8.6).

For shallow, large diameter beds, the distributor pressure drop should be between 50% and 100% of the bed pressure drop. When the ratio bed height/diameter exceeds 1, this percentage can drop to 10. For rectangular beds, the characteristic linear dimension should be read instead of diameter. It

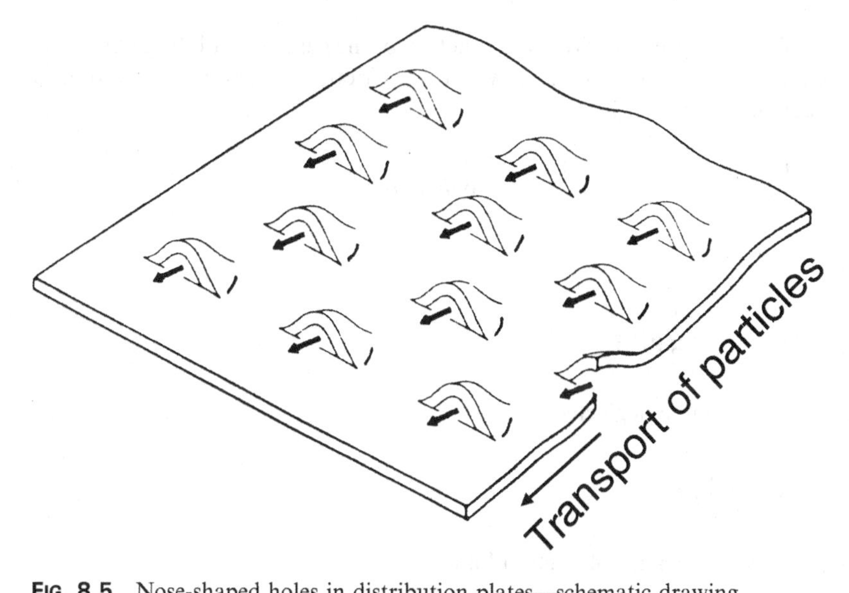

FIG. 8.5 Nose-shaped holes in distribution plates—schematic drawing.

is obtained by dividing the bed area by the perimeter (for more details, see Geldart and Baeyens, 1985). It is well known that if a gas distributor gives a pressure drop which is too low, the result is poor fluidization; that is, some parts of the bed will receive much less gas than others, and may be temporarily or permanently defluidized, while in other parts, the gas forms semipermanent spouts or channels.

FIG. 8.6 Nose-shaped holes in distribution plates. (Courtesy of Hein, Lehmann Trenn- und Fördertechnik, Düsseldorf, Germany.)

The plate pressure drop as a function of the gas flow and the geometry is calculated as follows. For a flow without friction, Bernoulli's equation is applicable (see Fig. 8.7).

$$\frac{1}{2}\rho v_1^2 + \rho g h + p_1 = \frac{1}{2}\rho v_2^2 + \rho g h + p_2$$

It follows that:

$$p_1 - p_2 = \frac{1}{2}\rho\left(v_2^2 - v_1^2\right)$$

Usually, v_1 can be neglected:

$$\Delta p_d = \frac{1}{2}\rho v_2^2$$

Δp_d is greater because of friction losses:

$$\Delta p_d = \frac{\rho v_2^2}{2C^2} \tag{8.3}$$

C is a friction factor.

Geldart and Baeyens (1985) provide information concerning the use of this equation for gas distributors of fluid beds. For the flow through a circular hole, the gas pressure and temperature are assumed to be equal to the gas temperature and pressure in the wind box.

Furthermore, for distributors having a free area of less than 10%, operating at an approach *Re* exceeding 3000 and with the ratio plate thickness/hole diameter smaller than 0.1, $C \approx 0.6$. For square-edged circular orifices

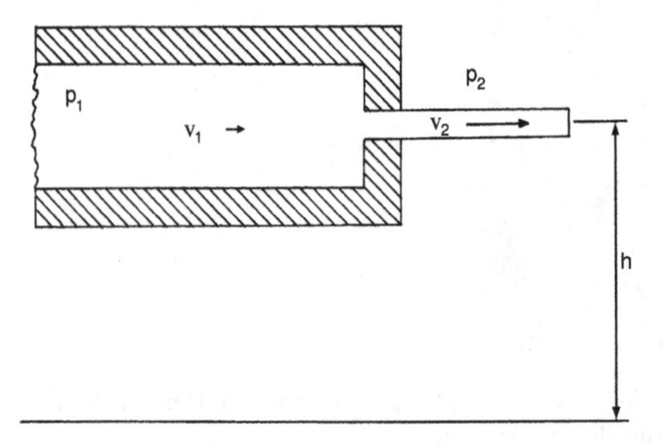

FIG. 8.7 Illustration for Bernoulli's equation for incompressible fluids.

with the ratio plate thickness/hole diameter exceeding 0.09, $C=0.82(t/d)^{0.13}$ is mentioned.

Example

Urea is cooled in a fluid bed cooler having a length of 12 m and a width of 3 m. In rest, the bed height is 0.3 m. The pressure in the disengagement section above the layer is atmospheric. The product's bulk density is 780 $kg \cdot m^{-3}$. A perforated plate is used having the following characteristics:

- Hole diameter 2 mm
- Plate thickness 2 mm
- 5% free area.

The airflow rate is 2.75 $kg \cdot m^{-2} \cdot sec^{-1}$.

Bed pressure loss: $780 \cdot 9.81 \cdot 0.3 = 2296\ N \cdot m^{-2}$.

Approach *Re*:

A characteristic linear dimension is found by dividing the bed area by the bed perimeter: $12 \cdot 3/(2 \cdot 12 + 2 \cdot 3) = 1.2$ m. In *Re*, four times this value is used: 4.8 m.

The air specific mass and viscosity are taken at atmospheric pressure and 20°C: 1.20 $kg \cdot m^{-3}$ and $1.6 \cdot 10^{-5}\ N \cdot sec \cdot m^{-2}$.

The linear gas velocity is $2.75/1.2 = 2.29\ m \cdot sec^{-1}$.

$$Re = \frac{1.20 \cdot 2.29 \cdot 4.8}{1.6 \cdot 10^{-5}} = 824,400$$

$Re > 3000$.

C:

The applicability of $C=0.82(t/d)^{0.13}$ is assumed.

It follows that: $C=0.82$.

Because the holes are nose-shaped, take $C = 0.9$.

The linear gas velocity in the holes is $2.29/0.05 = 45.8\ m \cdot sec^{-1}$.

Using Eq. (8.3):

$$\Delta p_d = \frac{1.2 \cdot 45.8^2}{2 \cdot 0.9^2} = 1554\ N \cdot m^{-2}$$

Ratio distributor pressure loss/bed pressure loss: $1554/2296 = 0.68$. The ratio bed height/bed diameter is $1/4.8 = 0.21$.

Notes:

The distributor pressure loss is strongly dependent on the airflow rate; the possibility of turndown must therefore be considered carefully when designing the distributor.

The air specific mass is actually slightly greater due to distributor and bed pressure losses.

8.7. PNEUMATIC NOZZLES

In pneumatic nozzles, the atomization is accomplished by the contact between a gas flow and a liquid flow. Usually, the gas flow is an air flow. The gas flows at a velocity in the range 100–200 $m \cdot sec^{-1}$, while the liquid flows at a velocity of several meters per second. The contact between the two flows causes the breakup of the continuous liquid flow. There are two types of pneumatic nozzles: with internal and external atomization. For AN and urea, pneumatic nozzles with external atomization are used only. Nozzle types with internal atomization can suffer from incrustations. Any pumpable feed can be atomized by the pneumatic nozzle. The maximum viscosity for a single-fluid nozzle is 0.4 $N \cdot sec \cdot m^{-2}$ (400 cP). For this reason and because of the possibility of plugging for single-fluid nozzles, pneumatic nozzles are preferred over single-fluid nozzles for viscous feeds. The tradeoff of this nozzle type is the power consumption. This will be illustrated by means of an example.

Only a fraction of the energy for passing the air through the nozzle is used for atomization.

Generally, three categories of pneumatic nozzles can be distinguished:

- Low pressure ($\leq$100 mm wg)
- Medium pressure (up to 0.7 bar)
- High pressure ($>$1 bar).

The maximum pressure drop of a pneumatic nozzle is 10 bar.

Pneumatic nozzles for fluid bed granulation are medium-pressure nozzles. Fig. 8.8 depicts a typical pneumatic nozzle for fluid bed granulation.

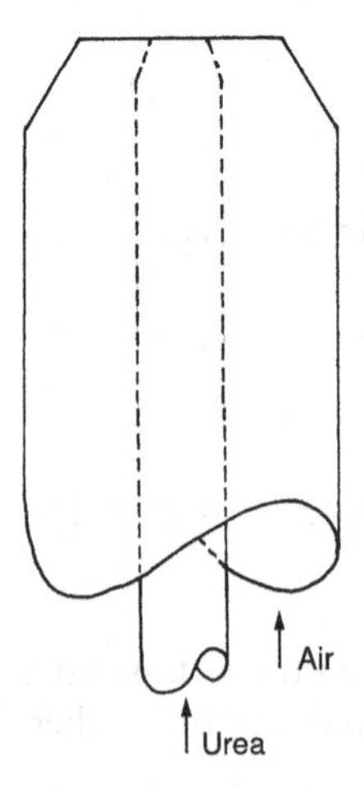

FIG. 8.8 A typical pneumatic nozzle for fluid bed granulation.

The air/liquid weight ratio of pneumatic nozzles is usually in the range 0.5–1.5. For fluid bed granulation, the ratio is close to 0.5. For a given nozzle design, the average droplet size of the spray produced is dependent on the gas/liquid ratio mainly. The average droplet size is also dependent on the air pressure drop. Typically, for water, the average droplet size varies from 100 to 40 μm when the air/liquid weight ratio varies from 0.4 to 1.0 and the air pressure loss is 4–5 bar (Lechler, 1999).

On granulating AN, the average spray droplet diameter is 0.1 mm (Bruynseels, 1985).

Example

A pneumatic nozzle has an external diameter of 22 mm. Two hundred normal cubic meters of air per hour are passed through the nozzle, while the air pressure loss is 0.5 bar. Four hundred kilograms of liquid AN per hour are atomized. The AN temperature is 150°C, while AN experiences a pressure loss of 0.1 bar. The AN specific mass is $1500\ \mathrm{kg \cdot m^{-3}}$.

It is required to calculate the power consumption of the nozzle in $\mathrm{kWhr \cdot t^{-1}}$ AN.

Due to the liquid flow:

Theoretical power requirement (incompressible fluid):

$$\phi_v \Delta p = \frac{400}{1500 \cdot 3600} \cdot 10{,}000 = 0.74\ \mathrm{W}$$

Actual power consumption will be approximately 1 W. The power consumption in $\mathrm{kWhr \cdot t^{-1}}$ due to the liquid flow is negligible.

Due to the airflow:

First, the theoretical energy per kilogram of air is calculated. The compression from atmospheric pressure to 1.5 bar is assumed to be adiabatic.

$$N_1 = \frac{\gamma}{\gamma - 1} \cdot \frac{RT_1}{M} \left[\left(\frac{p_2}{p_1} \right)^{(\gamma-1)/\gamma} - 1 \right] \quad \text{with } \gamma = \frac{c_p}{c_v}$$

$$N_1 = \frac{1.4}{1.4 - 1} \cdot \frac{8314 \cdot 293}{29} \left[\left(\frac{1.5}{1.0} \right)^{(1.4-1)/1.4} - 1 \right] = 36{,}110\ \mathrm{J \cdot kg^{-1}}$$

Second, the shaft power is calculated.

$$N_2 = \frac{\phi_m \cdot N_1}{\eta_{ad}} = \frac{200 \cdot 1.30 \cdot 36{,}110}{3600 \cdot 0.8} = 3260\ \mathrm{W}$$

η_{ad} is the efficiency of the adiabatic compression.

Power consumption in $\mathrm{kWhr \cdot t^{-1}}$ AN: $3.260/0.4 = 8.2$.

This power consumption is substantial.

The temperature of the air leaving the compressor can be calculated.

$$\frac{T_2}{T_1} = \left(\frac{p_2}{p_1}\right)^{(\gamma-1)/\gamma} = 1.123$$

$$T_2 = 1.123 \cdot 293 = 329 \text{ K}(56°\text{C})$$

After compression, the air temperature is raised to 150°C. Finally, it is possible to estimate the nozzle diameter of the AN line. Equation (8.3) is used:

$$\Delta p = \frac{\rho v^2}{2C^2} \text{ with } C = 0.8$$

It follows that: $v = 2.9 \text{ m} \cdot \text{sec}^{-1}$ and $d_h = 5.7$ mm.

8.8. THE HEAT BALANCE OF UREA GRANULATION

At the fluid bed granulation, heat is released by the cooling down of the melt and the melt crystallization. Heat is absorbed by the evaporation of water, the fluidization air, and the recycle. The atomization air is heated to slightly above the bed temperature to avoid precrystallization in the nozzles. At the crystallization of urea, it is not necessary to heat the ambient air used for fluidization. The fluidization air is preheated at the solidification of AN. The reason for this difference is a difference in the heats of crystallization of the two compounds. It is 224.5 $\text{kJ} \cdot \text{kg}^{-1}$ for urea, while it is 77.5 $\text{kJ} \cdot \text{kg}^{-1}$ for AN.

Example

Urea is crystallized in a fluid bed granulator having an area of 6 m^2. Thirty tons per hour of urea are produced. The bed temperature is 110°C. Per kilogram of urea produced, 0.5 kg is recycled. The recycle is at 60°C. In rest, the bed height is 1 m. The feed contains 96% urea by weight and is at a temperature of 160°C. The fluidization is accomplished by air having a temperature of 20°C. The air mass flow rate is 2 $\text{kg} \cdot \text{m}^{-2} \cdot \text{sec}^{-1}$. The water in the feed is evaporated. It is required to review the heat balance.

Urea data

T_o = 132.6°C
i = 224.5 $\text{kJ} \cdot \text{kg}^{-1}$
ρ_b = 740 $\text{kg} \cdot \text{m}^{-3}$
c_l = 2.10 $\text{kJ} \cdot \text{kg}^{-1} \cdot \text{K}^{-1}$
c_s = 1.75 $\text{kJ} \cdot \text{kg}^{-1} \cdot \text{K}^{-1}$ (25 – 132°C)

Air data

$\rho_g = 1.30\ \text{kg} \cdot \text{m}^{-3}$ at 0°C and atmospheric pressure
$c_g = 1.00\ \text{kJ} \cdot \text{kg}^{-1} \cdot \text{K}^{-1}$

Water data

$\Delta H_v = 2232\ \text{kJ} \cdot \text{kg}^{-1}$ (110°C)
$c_l = 4.2\ \text{kJ} \cdot \text{kg}^{-1} \cdot \text{K}^{-1}$

Urea and water mass balance ($\text{kg} \cdot \text{hr}^{-1}$)

	In	Out
Urea	30,000	30,000
Water	1250 +	0 +
	31,250	30,000

Granulation heat balance ($\text{kJ} \cdot \text{hr}^{-1}$)

Melt specific heat calculation

$$\frac{30{,}000}{31{,}250} \cdot 2.10 + \frac{1250}{31{,}250} \cdot 4.2 = 2.18\ \text{kJ} \cdot \text{kg}^{-1} \cdot \text{K}^{-1}$$

Heat release

Melt cooling	
31,250 · 2.18(160 − 110)	= 3,406,250
Urea crystallization	
30,000 · 224.5	= 6,735,000 +
	10,141,250

Heat absorption

Water evaporation	
1250 · 2232	= 2,790,000
Air heating	
6 · 2 · 1.00(110 − 20)3600	= 3,888,000
Recycle heating	
0.5 · 30,000 · 1.75(110 − 60)	= 1,312,500
Heat losses (balance)	= 2,150,750 +
	10,141,250

The bed holdup is $6 \cdot 1 \cdot 740 = 4440$ kg.
Calculation of the fluidization air velocity at 110°C and 1.1 bar:

$$2 \cdot \frac{1}{1.30} \cdot \frac{1}{1.1} \cdot \frac{273+110}{273} = 2 \text{ m} \cdot \text{sec}^{-1}$$

8.9. FLUID BED COOLING

Fig. 8.9 illustrates the starting point of an ideally mixed crossflow. At any position in the bed, the temperature of the product leaving the mixing cell equals the temperature of the leaving air. The average residence time of the product in the cell is longer than the average residence time of the air. The temperature of the product decreases, while the product flows from the product inlet to the product outlet. Thus the driving force for heat transfer decreases with the length traveled as the cooling air temperature is constant. In such a case, $\mathrm{d}(T_\mathrm{p}(x) - \mathrm{TA_{IN}})/\mathrm{d}x$ is proportional to $(T_\mathrm{p}(x) - \mathrm{TA_{IN}})$. On integrating the corresponding differential equation, an expression containing a natural logarithm is obtained. A heat balance for the mixing cell having a length of $\mathrm{d}x$:

$$\rho_\mathrm{g} v_\mathrm{F} B \cdot \mathrm{d}x \cdot c_\mathrm{g}(T_\mathrm{p}(x) - \mathrm{TA_{IN}}) = -\,\phi_\mathrm{p} c_\mathrm{s} \cdot \mathrm{d}T_\mathrm{p}(x)$$

Rearranging and integrating:

$$\int_{x=0}^{x=L} \mathrm{d}x = -\frac{\phi_\mathrm{p} c_\mathrm{s}}{\rho_\mathrm{g} v_\mathrm{F} B c_\mathrm{g}} \int_{T_\mathrm{p}(x=0)}^{T_\mathrm{p}(x=L)} \frac{\mathrm{d}(T_\mathrm{p}(x) - \mathrm{TA_{IN}})}{(T_\mathrm{p}(x) - \mathrm{TA_{IN}})}$$

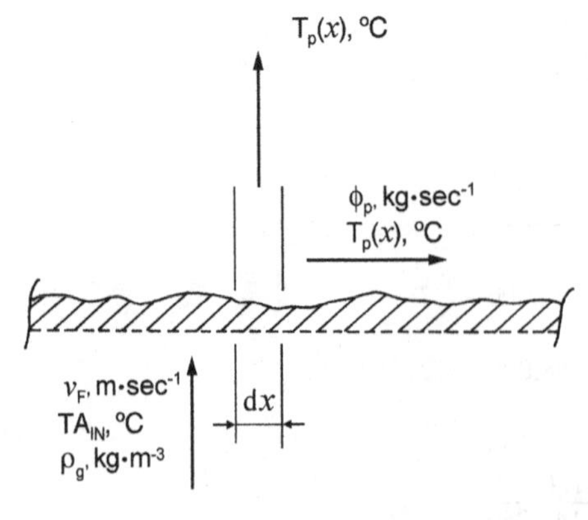

FIG. 8.9 Crossflow air heating of particulate material.

$$L = \frac{\phi_p c_s}{\rho_g v_F B c_g} \; {}^e\log \frac{T_p(x=0) - TA_{IN}}{T_p(x=L) - TA_{IN}} \; \text{m} \qquad (8.4)$$

Table 8.3 contains data concerning the applicability of the crossflow cooling model. Most of the data were collected within Akzo Nobel. Table 8.4 summarizes the physical properties of the materials mentioned in Table 8.3.

Biot's number provides the possibility of an assessment of the relative importance of the heat transfer to the particle and the heat dissipation in the particle itself.

$$Bi = \frac{\frac{\alpha_o}{\lambda_s}}{d_{50}/2}$$

The Biot number for the fluid bed cooling of CAN is estimated in Appendix A. The estimation shows that the resistance to heat transfer is located mainly in the boundary layer surrounding the particle. This means that there will be hardly a temperature gradient within the particle.

Furthermore, the estimation shows that the coefficient for the heat transfer from the particles to the air is high: 294 $W \cdot m^{-2} \cdot K^{-1}$. This figure can be compared to the range for the overall coefficients for the heat transfer from central heating radiators to air: 5–15 $W \cdot m^{-2} \cdot K^{-1}$.

The area offered by 100 kg of 3.3-mm spherical CAN particles is 101 m^2. The large bed area and the high heat transfer coefficient render the observations of Table 8.3 understandable. The Heat Atlas (1993) describes a more sophisticated model. In this source, the latter model is applied to the

TABLE 8.3 Applicability of the Model for Crossflow Cooling in a Plugflow Fluid Bed Cooler

Product	Scale of tests[a]	h_{exp} (mm)	v_F ($m \cdot sec^{-1}$)	TA_{IN} (°C)
Anhydrous citric acid	p	100	0.7/1	15
Citric acid monohydrate	p	70/200	1	15
CAN[b]	i	[c]	1.2	34
Cubic vacuum pan salt	i	200	0.9	26

[a] p: Pilot plant scale; i: industrial scale.
[b] Data from Winterstein et al. (1964).
[c] Not stated.

TABLE 8.4 Physical Properties of Fluid-Bed-Cooled Materials

Product	d_p (mm)	ρ_s ($kg \cdot m^{-3}$)	ρ_b ($kg \cdot m^{-3}$)	c_s ($kJ \cdot kg^{-1} \cdot K^{-1}$)	λ_s ($W \cdot m^{-1} \cdot K^{-1}$)
Anhydrous citric acid	0.45	1665	900	1.210	–
Citric acid monohydrate	0.65	1540	800	1.470	–
CAN	2[a]	1800[b]	1070	1.60[c]	2.0[d]
Cubic vacuum pan salt	0.4	2160	1250	0.870	5.8

[a] Quoted by Winterstein et al. (1964). [b]At 20°C. [c] Temperature range 32.3–84.2°C. [d]Osman et al. (1996).

batchwise fluid bed cooling of roasted coffee beans from 300°C to 35°C. There are 60 kg of beans per square meter. The specific air mass flow is $3\ kg \cdot m^{-2} \cdot sec^{-1}$ and the air is at 20°C.

Further data:

$d_p = 6$ mm
$\rho_s = 630\ kg \cdot m^{-3}$
$c_s = 1700\ J \cdot kg^{-1} \cdot K^{-1}$
$\lambda_s = 0.10\ W \cdot m^{-1} \cdot K^{-1}$

In the Heat Atlas, it is estimated that cooling for 150 sec will reduce the temperature of the beans to a temperature in the range 25.3–34.3°C. It is now calculated what the simple model would predict. However, Eq. (8.4) cannot be used directly because it was derived for continuous cooling. At batchwise cooling, the driving force for heat transfer decreases with the time elapsed. In such a case, $d(T_p(t) - TA_{IN})/dt$ is proportional to $(T_p(t) - TA_{IN})$. A heat balance for 1 m^2 of bed area in a small time dt:

$$\rho_g v_F \cdot 1 \cdot c_g \cdot dt(T_p(t) - TA_{IN}) = -Mc_s \cdot dT_p(t)$$

Rearranging and integrating:

$$\int_{t=0}^{t=t} dt = -\frac{Mc_s}{\rho_g v_F \cdot 1 \cdot c_g} \int_{T_p(t=0)}^{T_p(t=t)} \frac{d(T_p(t) - TA_{IN})}{(T_p(t) - TA_{IN})}$$

$$t = \frac{Mc_s}{\rho_g v_F \cdot 1 \cdot c_g} {}^{e}\log \frac{TA_{IN} - T_p(t=0)}{TA_{IN} - T_p(t=t)} \ \text{sec} \qquad (8.5)$$

This equation predicts cooling to 23.5°C for a cooling time of 150 sec:

$$150 = \frac{60 \cdot 1700}{3 \cdot 1000} \cdot {}^{e}\log \frac{20 - 300}{20 - 23.5}$$

Sixty kilograms of roasted coffee beans per square meter corresponds with an expanded bed height of 30–40 cm. Note that the diameter of the beans is relatively large (6 mm) and the thermal conductivity is low (0.10 $W \cdot m^{-1} \cdot K^{-1}$). Still, the simple model is approximately correct. The fluid bed coolers are operated with relatively shallow layers, e.g., 200-mm expanded bed height. First, this confers a plug flow character to the fluid bed cooler. Second, energy costs are saved because it requires less energy to fluidize a shallow layer than to fluidize a bed having an expanded height of, for example, 1 m.

Example

CAN, 100 $t \cdot hr^{-1}$, containing 27% nitrogen by weight are cooled from 115°C to 55°C in a fluid bed cooler by means of air having a temperature of 40°C. The expanded bed height is 200 mm. The pressure in the freeboard of the fluid bed cooler is atmospheric, while the pressure loss of the distributor and the fluidized bed is 200 mm wg. The gas velocity in the wind box is 3.5 $m \cdot sec^{-1}$. The bed width is 3 m. It is desired to calculate the required bed length.

AN data are taken instead of CAN data.

c_s : 1.90 $kJ \cdot kg^{-1} \cdot K^{-1}$ 84.2 – 125.9°C

1.60 $kJ \cdot kg^{-1} \cdot K^{-1}$ 55 – 84.2°C

Transition heat at 84.2°C: 17.0 $kJ \cdot kg^{-1}$

Air data:

$$\rho_g = 1.30 \cdot \frac{273}{273 + 40} \cdot \frac{10{,}526}{10{,}326} = 1.16\ kg \cdot m^{-3}$$

$$c_g = 1.00\ kJ \cdot kg^{-1} \cdot K^{-1}$$

The strategy is to carry out three calculations. First, the product is cooled from 115°C to 84.2°C. Second, at 84.2°C, a transition occurs. Seventeen kilojoules per kilogram are liberated. Third, the product is cooled from 84.2°C to 55°C.

$$L_1 = \frac{100{,}000 \cdot 1900}{3600 \cdot 1.16 \cdot 3.5 \cdot 3 \cdot 1000} {}^{e}\log \frac{115 - 40}{84.2 - 40} = 2.29\ m$$

$$L_2 = \frac{100{,}000 \cdot 17{,}000}{3600 \cdot 1.16 \cdot 3.5 \cdot 3 \cdot 1000(84.2 - 40)} = 0.88\ m$$

$$L_3 = \frac{100,000 \cdot 1600}{3600 \cdot 1.16 \cdot 3.5 \cdot 3 \cdot 1000} {}^{e}\log \frac{84.2 - 40}{55 - 40} = 3.94 \text{ m}$$

The three lengths add up to 7.11 m.

$$L_{\text{tot}} = 1.15 \cdot 7.11 = 8.18 \text{ m}$$

8.10. TEC GRANULATION PROCESS

TEC stands for Toyo Engineering Corporation. See Fig. 8.10 for a typical TEC plant for the granulation of urea. Morikawa and Kido (1998) have described such a plant built at Priesteritz in Germany. The plant has a capacity of 1200 t per day. It resembles Hydro Agri's granulation plant as described in Sec. 8.2. However, there are differences concerning the design. The Spout–Fluid Bed Granulator is the heart of the plant (see Fig. 8.11). The feed solution, containing minimum 95% urea by weight and having a temperature of, for example, 135°C is atomized upwards into a bed of moving particles by means of single-fluid nozzles located in a spout pipe. Spouting air and fluidizing air are introduced into the bed separately. Neither the fluidization air nor the spouting air is heated. The fluidization air is evenly distributed by a perforated distributor plate.

The pressure loss of both spouting air and fluidizing air is in hundreds of mm wg. The bed temperature is 100–105°C. The underpressure in the granulator is −10 to −20 mm wg. The heat of solidification of urea is removed

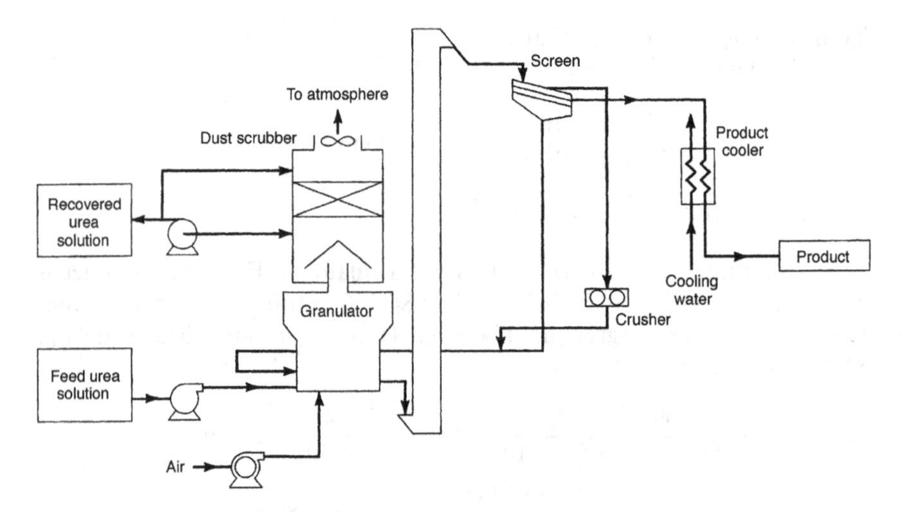

FIG. 8.10 A typical TEC granulation plant for urea. (Courtesy of Toyo Engineering Corporation, Chiba, Japan.)

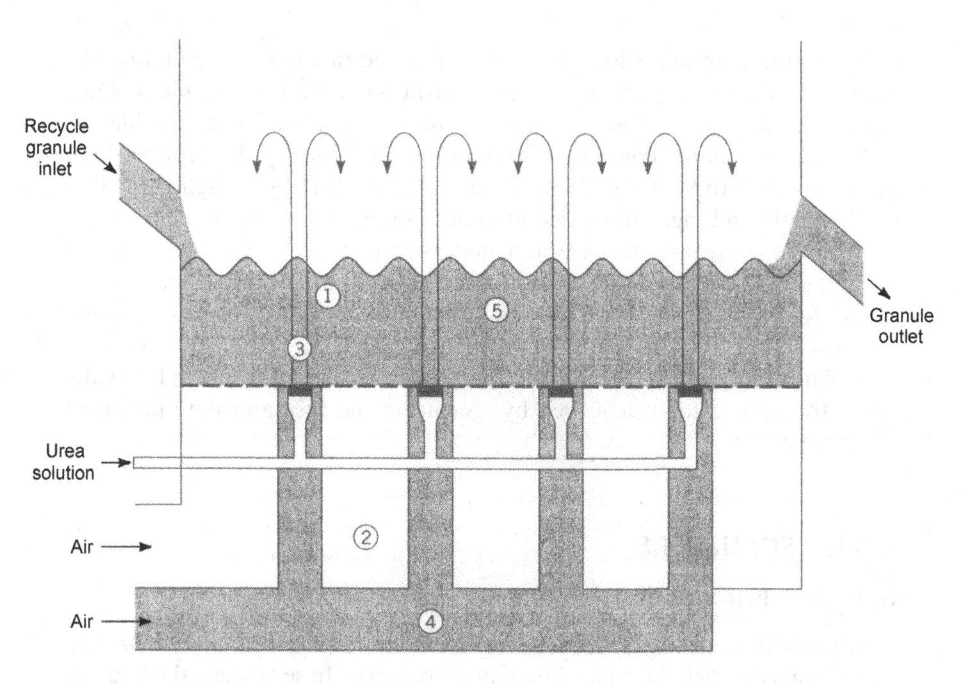

FIG. 8.11 Spout-Fluid Bed Granulator. 1: spouted bed; 2: plenum; 3: spray nozzle; 4: air header; 5: fluidized bed. (Courtesy of Toyo Engineering Corporation, Chiba, Japan.)

by the air and the evaporation of water in the feed solution. The fluid bed granulator is of the rectangular type; that is, the length is 3 to 4 times larger than the width. The material travels more or less in plug flow through the granulator.

Per kilogram of urea produced by the plant, 0.5–0.7 kg of fine and crushed particles are fed to the granulator. The crushed feed is $2.3\ t \cdot hr^{-1}$, i.e., 4–6% of the urea fed to the granulation plant. The number of fresh seed particles per unit of time (fine and crushed material) almost equals the number of particles produced per unit of time by the granulator. Between the inlet and the outlet of the granulator, there are four parallel rows of spout pipes. There are nine spout pipes and nozzles per row of pipes. Thirty-two out of the 36 nozzles are active, while the last 4 nozzles are standby.

Formaldehyde is added to the feed urea solution to reduce dust formation in the granulator and to improve crushing strength and flow properties of the particles (see Sec. 8.2 for a detailed description of this effect).

The air leaving the granulator contains some rather coarse urea dust which is caught in a packed-bed scrubber. The liquid sprayed into the scrubber is an aqueous solution containing 30% urea by weight (see Sec. 8.11). The dust formation amounts to 2–3% of the urea feed rate to the granulator. The recovered urea is recycled to the concentration section of the plant. Dust outlet concentrations of less than 30 $mg \cdot nm^{-3}$ in the stack can be achieved.

The granulator contains a cooling section bringing the product temperature down to 90–95°C. Now the granules are strong enough to survive mechanical handling. The cooled granules are separated into three fractions by means of screens. Oversize and undersize products are recycled to the granulator. The coarse fraction is crushed after it has left the screen.

The on-spec product from the screens is cooled by means of cooling water in contact coolers. The urea temperature is reduced from 90°C to 50°C.

The granulator is cleaned once per 2 months. Cleaning lasts 8 hr. Scale-up of the granulator is achieved by increasing the area and the number of spout pipes.

8.11. SCRUBBERS

8.11.1. Introduction

Scrubbers form a class of devices in which a liquid (usually water) is used to assist or accomplish the collection of dusts or mists. In scrubbers, the liquid is dispersed into the gas as a spray and the liquid droplets are the principal collectors for the dust particles. The principles of the collection of particles are illustrated in Fig. 8.12.

First, inertial interception may be distinguished. A particle carried along by the gas stream tends to follow the stream but may strike a fiber or liquid droplet because of its inertia. The fluid streamlines are represented by solid lines, and the dotted lines represent the paths of particles that initially followed the fluid streamlines. P and Q are the limiting streamlines.

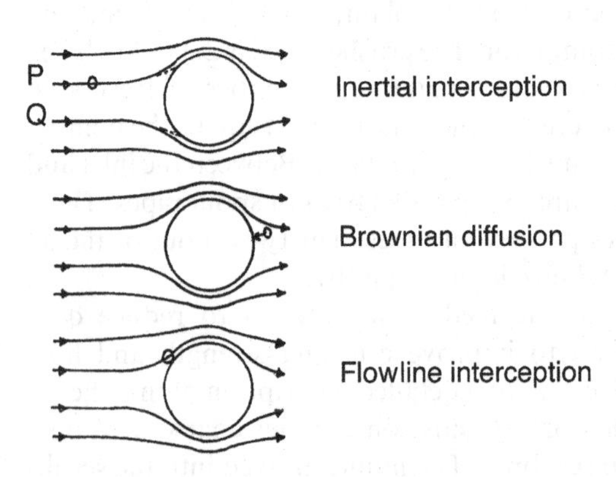

FIG. 8.12 Particle deposition in scrubbers.

The second mechanism is Brownian diffusion. Small particles, particularly those below about 0.3 μm in diameter, do not move uniformly along the gas streamline. They diffuse from the gas to the surface of the collecting body. The Brownian movement is a result of collisions between the small particles and gas molecules.

The third deposition method is flowline interception. A particle traveling along a streamline will touch the body if the streamline passes the collecting body within one particle radius.

The predominant mechanism is inertial interception. Scrubbers are effective for particles larger than 0.1 μm.

Two functions can be distinguished in a scrubber:

- A contacting function, in which a spray is generated and the dust-laden gas stream is brought into contact with it.
- A separation function, in which the spray and deposited dust particles are separated from the cleaned gas.

The two stages may be separate or physically combined. In the contactor stage, effective contacting of the gas and spray is brought about. There are three methods to generate the spray:

- By means of single-fluid or pneumatic nozzles
- By the flow of the gas itself
- By a motor-driven rotor through which both gas and liquid pass.

It is customary to term devices in these classes:

- Spray scrubbers
- Gas-atomized spray scrubbers
- Mechanical scrubbers.

Entrainment separation is accomplished with inertial separators, which are usually cyclones or impingement separators. The efficiency of a scrubber depends on essentially total removal of the spray from the gas stream.

Generally, the sprays generated in scrubbers are large enough in droplet size that they can be readily removed by properly designed inertial separators. However, re-entrainment is a usual problem. This is often caused by too high gas velocities. Furthermore, the entrainment separator should not readily be blocked by solids, and it should be possible to clear it of deposits should they occur. Cyclone separators are advantageous in this respect and are widely used with venturi contactors. However, they cannot readily be combined with other scrubber types, which can be more conveniently fitted with various forms of impingement separators. The diversity of scrubber designs is great. It is possible to distinguish four successful scrubber types. Three of these designs are shown schematically in Fig. 8.13. The fourth scrubber type is

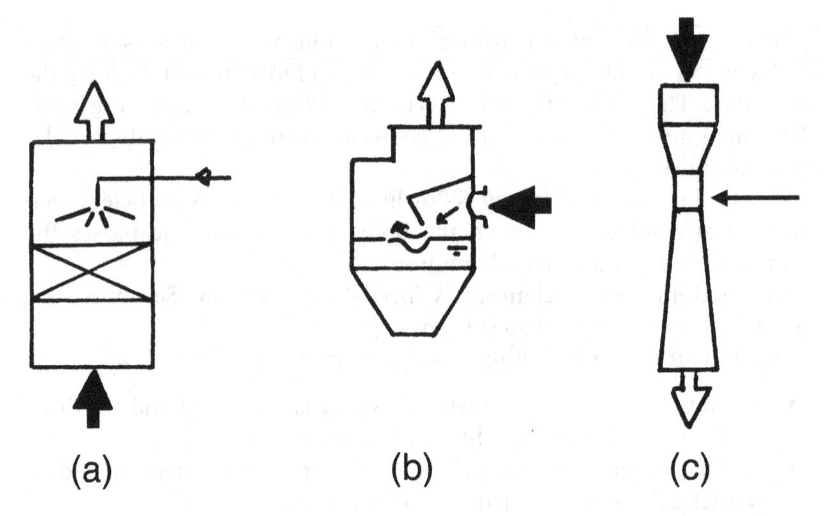

FIG. 8.13 Three scrubber designs. (a) Packed-bed scrubber, (b) self-induced spray scrubber, (c) venturi scrubber.

shown in Figs. 8.14–8.16. The four types are discussed shortly. A start is made with the scrubbers in Fig. 8.13.

The gas to be cleaned flows from the bottom to the top through a packed-bed scrubber. A packed-bed scrubber is a spray scrubber. Water is sprayed onto the packings. Entrained droplets are separated by means of an impingement separator. Packed-bed scrubbers can be used for water-soluble solids. They are effective for particles larger than 1 μm.

In a self-induced spray scrubber, the gas stream flows at high velocity through the contactor and atomizes the cleaning liquid. A self-induced spray scrubber is a gas-atomized spray scrubber. Any entrained droplets settle in the freeboard. This scrubber is also effective for particles smaller than 1 μm. The scrubbing efficiency decreases when the gas flow decreases.

The venturi scrubber is one of the most widely used types of particulate scrubbers. This scrubber also belongs to the category of the gas-atomized spray scrubbers. The gas velocity is accelerated in the throat and the scrubbing liquid is dosed upstream of the throat. Most commonly, venturi scrubbers are employed as high-energy units. They are effective for particles larger than 0.1 μm. The scrubbing efficiency also decreases when the gas flow decreases. The entrainment separator is commonly of the cyclone type.

In Figs. 8.14–8.16, a mechanical scrubber is depicted. In Fig. 8.14, the dust-laden gas enters at the left. The gas then flows through the scrubber, which contains a rotor and a stator. Both the rotor and the stator are equipped with bars (see Fig. 8.15). Water is distributed to the scrubber. The

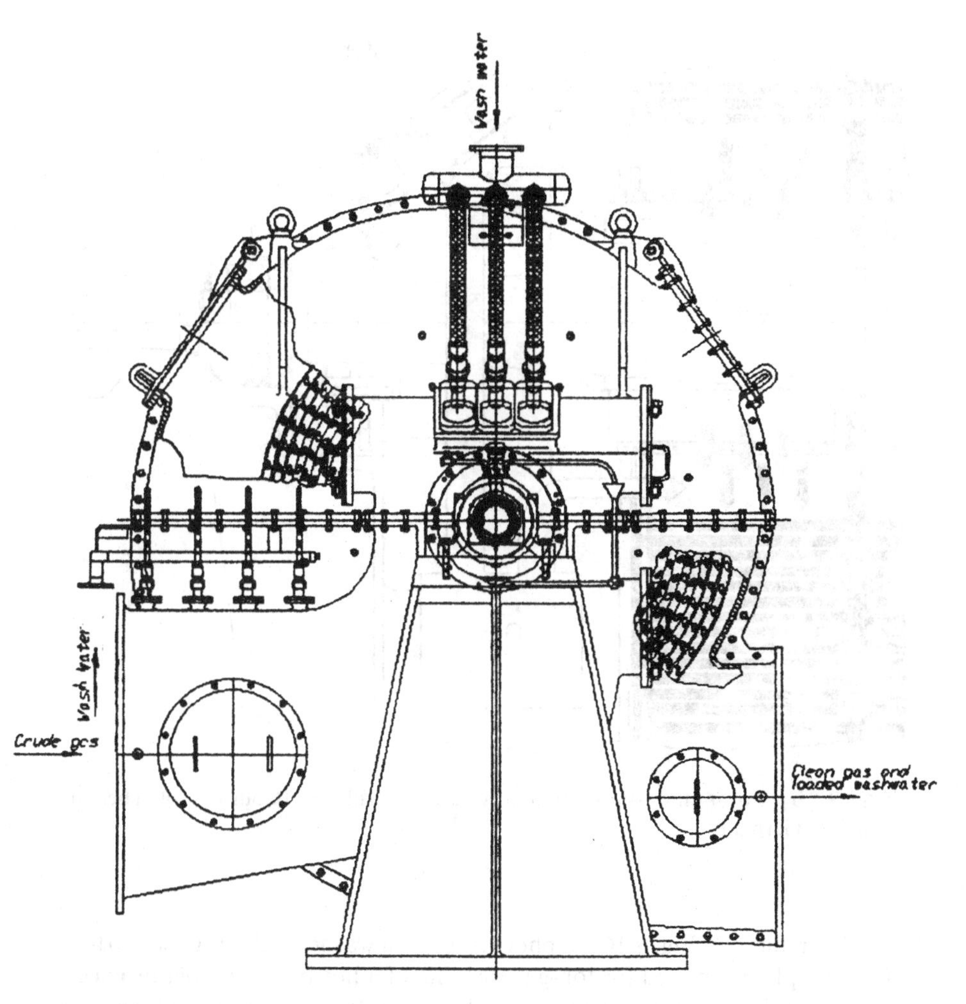

FIG. 8.14 Mechanical scrubber. (Courtesy of Theisen, Munich, Germany.)

rotating and stationary bars atomize the water thoroughly. The water droplets capture the dust, and liquid/gas separation is the next step. Particles as small as 0.1 μm can be removed effectively. The energy consumption can be as high as 15 kWhr per 1000 nm^3 gas. These pieces of equipment are widely used in the metallurgical industries where they are known as "Theisen washers." Due to their design features, an additional fan is not required because these scrubbers incorporate two functions in one piece of equipment:

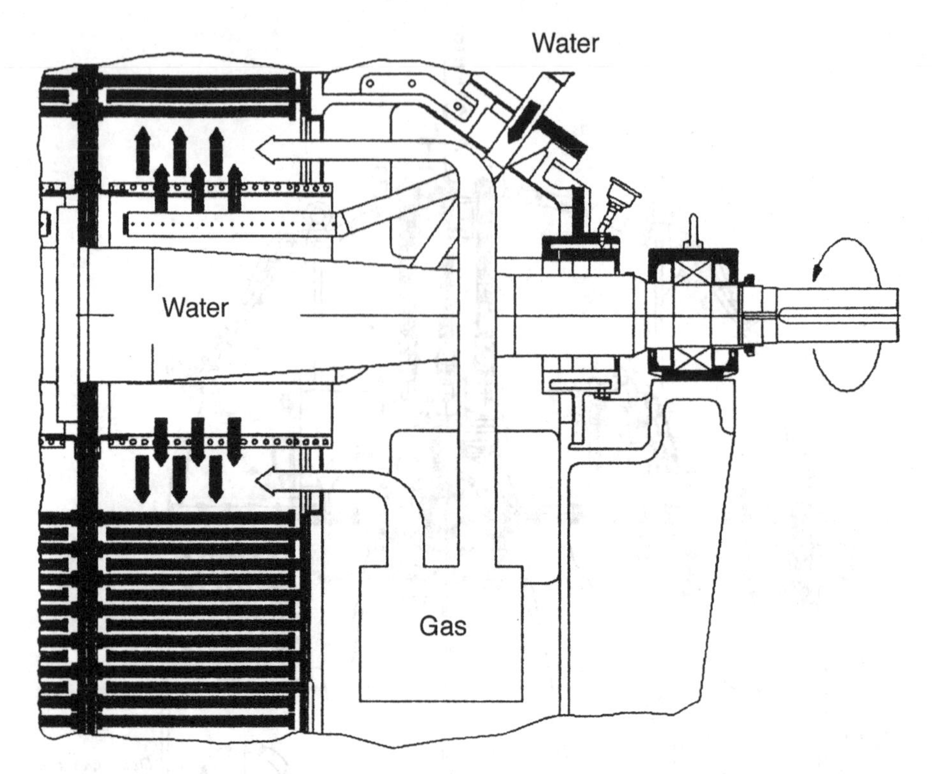

FIG. 8.15 Rotor and stator of a mechanical scrubber. (Courtesy of Theisen, Munich, Germany.)

scrubber and fan. Fig. 8.16 is a photograph of a washer including the start-up fluid coupling supplied for top-gas cleaning of a hot-blast coupola furnace.

Fans into which water is sprayed are simple mechanical scrubbers but do not correspond to the effectivity in dust removal required today.

The collection efficiency of a scrubber is essentially dependent only on the contacting power and is affected to only a minor degree by the size or geometry of the scrubber or by the way in which the contacting power is applied (Perry, 1997). Contacting power is defined as the energy per unit mass or volume of the gas. The energy supplied is dissipated in gas/liquid contacting and is ultimately converted into heat.

It has sometimes been stated that multiple gas/liquid contactors in series will give higher efficiencies at a given contacting power than will a single contacting stage. There is, however, little experimental evidence to support this contention (Perry, 1997).

FIG. 8.16 Mechanical scrubber. (Courtesy of Theisen, Munich, Germany.)

8.11.2. A Typical Scrubber for a Fluid Bed Granulation Plant

A typical scrubber for a fluid bed granulation plant is depicted in Fig. 8.17. Koch-Glitsch manufactures these scrubbers and calls them MultiVenturi FlexiTray scrubbers (MVFT). As shown in this figure, the scrubber comprises one or more trays. Each tray contains numerous venturi openings. Each of the MultiVenturi openings is surmounted by a spider cage holding a floating Flexicap (see Fig. 8.18). In addition, each tray is equipped with one or more downcomers and weir flow baffles that control the scrubbing liquid as it flows across the tray and then to the tray below.

The vapor or gas enters the bottom inlet and flows upward through the caps. At low gas velocities, the lightweight caps (located in every other row) rise first, whereas the heavyweight caps (in the alternate rows) remain in the "closed" position. Actually, the tabs on the caps always allow some gas leakage to prevent the caps from sticking to the deck. All the caps are fully opened as the vapor flow attains the design conditions.

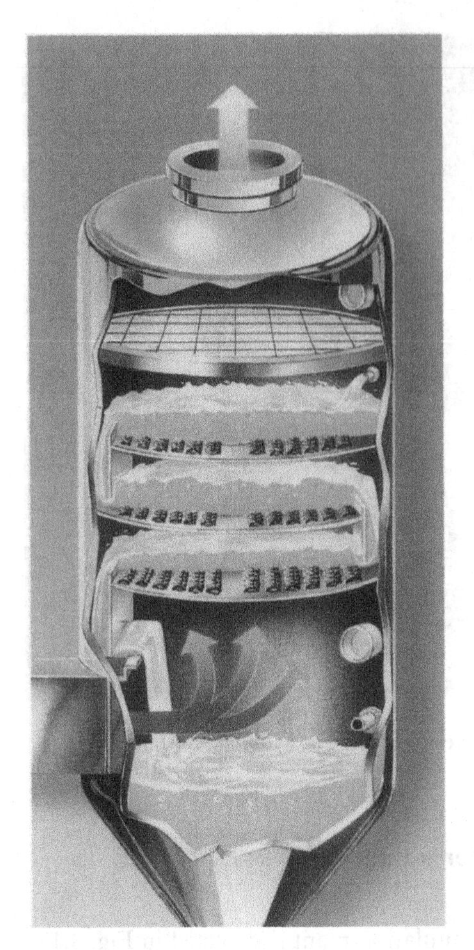

FIG. 8.17 Typical scrubber for a fluid bed granulation plant. (Courtesy of Koch-Glitsch Italia, Milan, Italy.)

The liquid flows across the deck and is kept in a constant froth by the gas which exits each cap at "venturi velocity." There is always a head of frothy liquid maintained by the weir. As a result, complete gas and liquid contact is assured. The MultiVenturi FlexiTray, like the conventional venturi, divides the liquid flow into fine droplets that capture particulates.

Before the gas leaves the scrubbing chamber, it passes through a high-efficiency mist eliminator. Entrainment-free gas exits from the top of the unit.

In addition to the aforementioned text, several remarks are made.

The MVFT is a variable-orifice design, the orifice opening being regulated only by the energy of the gas it is scrubbing. As a result, essentially

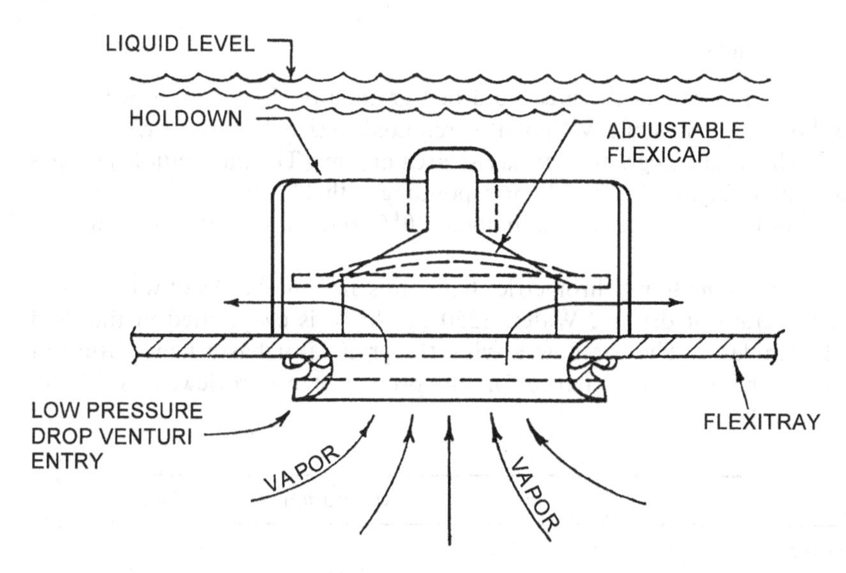

FIG. 8.18 MultiVenturi opening. (Courtesy of Koch-Glitsch Italia, Milan, Italy.)

constant pressure drop is achieved over a wide range of vapor flow rates: 30–110% of design gas flow.

The pressure drop of a tray is in the range 25–50 mm wg. Turbulence from closely spaced venturis prevents settling and "plating out" of solids. Scale-up is accomplished by increasing cross-sectional areas. In this way, a large unit has the same size venturi opening but more of them.

MultiVenturi Sizing Chart (Courtesy of Koch-Glitsch Italia, Milan, Italy)

Saturated gas volume ($m^3 \cdot hr^{-1}$)	Scrubber diameter (mm)	Overall scrubber height (mm)
4000	850	4350
8500	1150	4650
17,000	1500	4850
42,500	2350	5750
85,000	3200	6850
125,000	3950	7350
170,000	4500	7900
250,000	5300	8900
350,000	6250	9650
500,000	7500	10,700
850,000	9500	12,300

Example

The air leaving the urea fluid bed granulator described in Sec. 8.8 is scrubbed by means of a MVFT. It is required to size the scrubber.

The fluid bed granulator has an area of 6 m^2. The fluidization air mass flow rate is 2 $kg \cdot m^{-2} \cdot sec^{-1}$ corresponding with 43,200 $kg \cdot hr^{-1}$.

Assumption: the air is initially at 20°C and has a relative humidity of 60%.

According to psychrometric charts, this air contains 9 g of water vapor per kilogram of dry air. Water, 1250 $kg \cdot hr^{-1}$, is evaporated in the fluid bed granulator. The gas flow leaving the granulator has a temperature of 110°C. The gas mass balance for this process step is reviewed as follows ($kg \cdot hr^{-1}$):

	In	Evaporation	Out
Dry air	42,815	0	42,815
Water vapor	385 +	1250 +	1635 +
	43,200	1250	44,450
Temperature (°C)	20	110	110
Gram of water per kilogram of dry air	9	–	38

The air having a temperature of 110°C and containing 38 g of water vapor per kilogram of dry air enters the scrubber. In the scrubber, the gas flow is contacted with an aqueous urea solution containing 45% urea by weight.

The gas flow is cooled by evaporation. A scrubber temperature of 49°C is assumed and this assumption is checked as follows. Raoult's law says that the saturated water vapor pressure of a 45% solution equals the water mole fraction in the solution times the saturated vapor pressure of pure water.

	g	$g \cdot mol^{-1}$	mol	Mole fraction
Urea	45	60	0.75	0.20
Water	55 +	18	3.06 +	0.80 +
	100		3.81	1.00

Saturated water vapor pressure of a 45% solution: $0.80 \cdot 11{,}732 = 9386$ $N \cdot m^{-2}$.

The atmospheric pressure is 101,300 $N \cdot m^{-2}$. Thus the mole fraction of water vapor in the gas flow leaving the scrubber is 9386/101,300 = 0.093.

The weight fraction of water vapor in the gas flow leaving the scrubber is

$$\frac{0.093 \cdot 18}{0.093 \cdot 18 + 0.907 \cdot 29} = 0.060$$

So the water vapor flow leaving the scrubber is

$$\frac{0.06}{0.94} \cdot 42{,}815 = 2733 \text{ kg} \cdot \text{hr}^{-1}$$

and the evaporation of water in the scrubber is

$$2733 - 1635 = 1098 \text{ kg} \cdot \text{hr}^{-1}.$$

The heat flow needed to evaporate this water is

$$1098 \cdot 2386 = 2{,}619{,}828 \text{ kJ} \cdot \text{hr}^{-1}.$$

The heat of evaporation at 49 °C is 2386 $kJ \cdot kg^{-1}$.

The heat flow delivered by the cooling gas is

$$44{,}450 \cdot 1(110 - 49) = 2{,}711{,}450 \text{ kJ} \cdot \text{hr}^{-1}$$

The air specific heat in $kJ \cdot kg^{-1} \cdot K^{-1}$ is 1.

The two heat flows are approximately equal and a scrubber temperature of 49 °C is a correct assumption.

	$kg \cdot hr^{-1}$	$kg \cdot m^{-3}$	$m^3 \cdot hr^{-1}$
Dry air	42,815	1.102	38,852
Water vapor	2733 +	0.684	3996 +
	45,548		42,848

The saturated gas volume in $m^3 \cdot hr^{-1}$ is 42,848. According to the Multi-Venturi Sizing Chart, a scrubber having a diameter of 2.5 m and a height of 6 m is selected.

The scrubber is equipped with three trays.
Concerning the pressure drop, Fig. 8.19 is used.
The saturated gas velocity is 8.0 $ft \cdot sec^{-1}$.
The saturated gas density is 0.066 $lb \cdot ft^{-3}$.

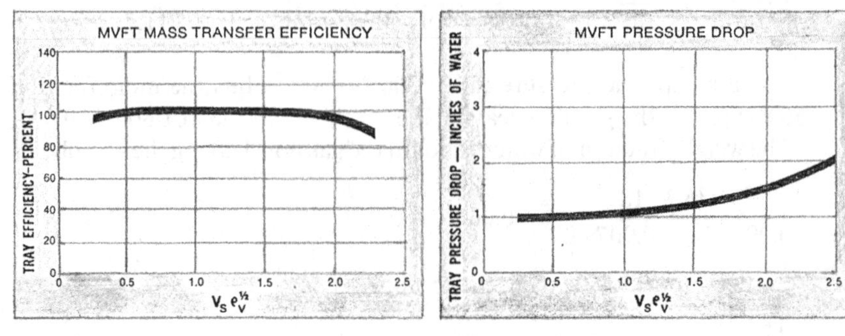

FIG. 8.19 MultiVenturi FlexiTray mass transfer efficiency and pressure drop. (Courtesy of Koch-Glitsch Italia, Milan, Italy.)

Tray pressure drop 1.5 in. of water.

Assumption: demister pressure drop also 1.5 in. of water.

Scrubber pressure drop: $4 \cdot 1.5 \cdot 25.4 = 152.4$ mm wg (152.4 mm wg corresponds with $1495.0\ \mathrm{N \cdot m^{-2}}$).

$$\text{Fan power consumption}: \frac{42,848 \cdot 1495.0}{3600 \cdot 1000 \cdot 0.5} = 35.6\ \text{kW}.$$

Overall efficiency: 0.5.

A 50-kW motor is selected for the fan.

$$\frac{35.6}{30} = 1.20\ \text{kWhr} \cdot \text{t}^{-1}\ \text{of urea}$$

One kilowatthour per 1000 $\mathrm{nm^3}$ of gas.

LIST OF SYMBOLS

Ar	Archimedes number $\left[\frac{g d_p^3 \rho_g (\rho_s - \rho_g)}{\mu_g^2}\right]$
a	Thermal diffusion coefficient $\{\lambda/(c_p \rho)\}$ $[\mathrm{m^2 \cdot sec^{-1}}]$
B	Fluid bed width (m)
Bi	Biot number $\{\alpha_o d_{50}/(2\lambda_s)\}$
C	Friction factor
c_g	Gas specific heat $[\mathrm{J \cdot kg^{-1} \cdot K^{-1}}]$
c_l	Melt specific heat $[\mathrm{J \cdot kg^{-1} \cdot K^{-1}}]$
c_p	Gas specific heat at constant pressure $[\mathrm{J \cdot kg^{-1} \cdot K^{-1}}]$
c_s	Solid specific heat $[\mathrm{J \cdot kg^{-1} \cdot K^{-1}}]$
c_v	Gas specific heat at constant volume $[\mathrm{J \cdot kg^{-1} \cdot K^{-1}}]$
c_w	Resistance coefficient (falling particle)

d	Fluid bed distributor hole diameter [m]
	Linear dimension [m]
d_h	Nozzle diameter [m]
d_p	Particle diameter [m]
$\bar{d}_p$	Particle Sauter mean diameter [m]
d_{50}	Average particle size (weight basis) [m]
G_{mf}	Minimum fluidization mass velocity $[kg \cdot m^{-2} \cdot sec^{-1}]$
g	Acceleration due to gravity $[m^2 \cdot sec^{-1}]$
ΔH_v	Water heat of evaporation $[J \cdot kg^{-1}]$
h	Fluid bed layer height (settled) [m]
	Height [m]
h_{exp}	Experimental fluid bed layer height [m]
i	Heat of fusion $[J \cdot kg^{-1}]$
k_f	Form factor
L	Fluid bed cooler length [m]
L_{tot}	Total CAN fluid bed cooler length [m]
L_1	CAN fluid bed cooler length—first part [m]
L_2	CAN fluid bed cooler length—second part [m]
L_3	CAN fluid bed cooler length—third part [m]
M	Molecular weight $[kg \cdot kmol^{-1}]$
	Mass [kg]
Nu	Nusselt number $(\alpha_o d_p/\lambda_g)$
Nu_{lam}	Nusselt number for laminar flow
Nu_{turb}	Nusselt number for turbulent flow
N_1	Theoretical air compression energy $[J \cdot kg^{-1}]$
N_2	Air compressor shaft power [W]
n	Number of particles in a size range
Pr	Prandtl number (ν/a)
p_1	Pressure upstream a nozzle or compressor $[N \cdot m^{-2}]$
p_2	Pressure downstream a nozzle or compressor $[N \cdot m^{-2}]$
Δp	Fluid bed or nozzle pressure loss $[N \cdot m^{-2}]$
Δp_d	Fluid bed distributor pressure loss $[N \cdot m^{-2}]$
R	Universal gas constant $[J \cdot kmol^{-1} \cdot K^{-1}]$
Re	Reynolds number $((vd)/\nu)$
TA_{IN}	Air inlet temperature [°C]
T_P	Product temperature [°C]
T_o	Melting point [°C]
T_1	Compressor air inlet temperature [K]
T_2	Compressor adiabatic air outlet temperature [K]
t	Plate thickness [m]
	Time [sec]
v	Gas velocity $[m \cdot sec^{-1}]$
v_F	Fluidizing medium velocity $[m \cdot sec^{-1}]$

v_{maxf} Maximum fluidization velocity [$m \cdot sec^{-1}$]
v_{mf} Minimum fluidization velocity [$m \cdot sec^{-1}$]
$v_{mf'}$ Minimum fluidization velocity below distributor plate [$m \cdot sec^{-1}$]
v_p Particle terminal velocity [$m \cdot sec^{-1}$]
v_s Saturated gas velocity [$ft \cdot sec^{-1}$]
v_1 Velocity upstream a nozzle [$m \cdot sec^{-1}$]
v_2 Velocity in a nozzle [$m \cdot sec^{-1}$]
X Weight fraction in a size range
x Length coordinate [m]
α_o Heat transfer coefficient [$W \cdot m^{-2} \cdot K^{-1}$]
γ c_p/c_v
η_{ad} Efficiency of the adiabatic compression
λ Thermal conductivity [$W \cdot m^{-1} \cdot K^{-1}$]
λ_g Gas thermal conductivity [$W \cdot m^{-1} \cdot K^{-1}$]
λ_s Solid thermal conductivity [$W \cdot m^{-1} \cdot K^{-1}$]
μ_g Gas dynamic viscosity [$N \cdot sec \cdot m^{-2}$]
ν Kinematic viscosity [$m^2 \cdot sec^{-1}$]
ρ Specific mass [$kg \cdot m^{-3}$]
ρ_b Bulk density [$kg \cdot m^{-3}$]
ρ_F Fluidizing medium specific mass [$kg \cdot m^{-3}$]
ρ_g Gas specific mass [$kg \cdot m^{-3}$]
ρ_s Solid specific mass [$kg \cdot m^{-3}$]
ρ_v Saturated gas density [$lb \cdot ft^{-3}$]
ϕ_m Mass flow [$kg \cdot sec^{-1}$]
ϕ_p Fluid bed granulator production [$kg \cdot sec^{-1}$]
ϕ_v Volumetric flow [$m^3 \cdot sec^{-1}$]

APPENDIX A

Calculation of the Biot Number of a CAN particle in a fluid bed cooler

Calcium ammonium nitrate physical properties

$d_p = 3.3 \cdot 10^{-3}$ m
$\rho_s = 1800\ kg \cdot m^{-3}$ (approximately)
$c_s = 1900\ J \cdot kg^{-1} \cdot K^{-1}$ (for AN in the range 84.2–125.9°C)
$\lambda_s = 2.0\ W \cdot m^{-1} \cdot K^{-1}$ (Osman et al., 1996)

Air physical properties (70°C)

$\rho_g = 1.03\ kg \cdot m^{-3}$
$c_g = 1010\ J \cdot kg^{-1} \cdot K^{-1}$
$\mu_g = 2.05 \cdot 10^{-5}\ N \cdot sec \cdot m^{-2}$
$\lambda_g = 0.029\ W \cdot m^{-1} \cdot K^{-1}$

The product is at 100°C while the air heats up from 40°C to 100°C.

The Heat Atlas (1993) approach is followed to calculate α_o, the heat transfer coefficient. In Sec. 5.5, the Frössling equation was discussed. This approach is not used here because *Re* exceeds 800. According to the Heat Atlas, the coefficient for the heat transfer from the particles in a fluidized bed to the gas is approximately equal to the heat transfer coefficient of a particle falling at terminal velocity.

The heat transfer through the laminary boundary layer is distinguished from the heat transfer in the turbulent wake. By using *Ar*, the Archimedes number, the calculation of the terminal velocity can be avoided. The Archimedes number is also known as the Grashof number.

$$Ar = g d_p^3 \rho_g (\rho_s - \rho_g)/\mu_g^2 = 1{,}554{,}405$$

$$Re = 18\left(\sqrt{1 + \frac{1}{9}\sqrt{Ar}} - 1\right)^2 = 2104$$

$$Pr = \frac{\nu}{a} = \frac{\mu_g c_g}{\lambda_g} = 0.71$$

$$Nu_{\mathrm{lam}} = 0.664 \cdot Re^{1/2} \cdot Pr^{1/3} = 27.2$$

$$Nu_{\mathrm{turb}} = \frac{0.037 \cdot Re^{0.8} \cdot Pr}{1 + 2.443 \cdot Re^{-0.1}(Pr^{2/3} - 1)} = 15.6$$

$$Nu = 2 + \sqrt{27.2^2 + 15.6^2} = 33.4$$

$$\alpha_o = \frac{\lambda_g}{d_p} Nu = 294\ \mathrm{W \cdot m^{-2} \cdot K^{-1}}$$

$$Bi = \frac{\alpha_o d_{50}}{2\lambda_s} = 0.24$$

Conclusion: The resistance to heat transfer is located mainly in the boundary layer surrounding the particle.

Note: A lower λ_s would have shown that the heat conduction in the particle cannot be neglected.

REFERENCES

Bruynseels, J.P. (1985). Fluid bed granulation of ammonium nitrate and calcium ammonium nitrate. *Proceedings of the Fertiliser Society of London, No. 235*. European Economic Community (1980). *E.E.C. Directive 80/876*. Brussels, Belgium.

Geldart, D., Baeyens, J. (1985). The design of distributors for gas fluidized beds. *Powder Technol.* 42:67.

Lechler, Germany. (1999). Private communication.

Leva, M. (1959). *Fluidization*. New York: McGraw-Hill.

Morikawa, H., Kido, K. (1998). TEC urea granulation technology—its salient features and industrial experiences. *Fertil. Ind.* 20:41.

Osman, M.B.S., Dakroury, A.Z., Dessouky, M.T., Kenawy, M.A., El-Sharkawy, A.A. (1996). Measurement of thermophysical properties of ammonium salts in the solid and molten states. *J. Therm. Anal.* 46:1697.

Perry, R.H., Green, D.W. (1997). *Perry's Chemical Engineers' Handbook*. New York: McGraw-Hill.

VDI (1993). *VDI Heat Atlas*. Düsseldorf, Germany: VDI Verlag.

Winterstein, G., Rose, K., Viehweg, H., Schreyer, L. (1964). Cooling of particulate materials in a rectangular fluid bed. *Chem. Tech.* 16:106. In German.

Index

Printed in the United States
by Baker & Taylor Publisher Services